AF335396

MECHANISMS OF CORTICAL INHIBITION

MECHANISMS OF CORTICAL INHIBITION

V.M. Okujava

NOVA SCIENCE PUBLISHERS, INC.

COMMACK, NEW YORK

Art Director: Maria Ester Hawrys
Assistant Director: Elenor Kallberg
Graphics: Susan A. Boriotti and Frank Grucci
Manuscript Coordinator: Phyllis Gaynor
Book Production: Gavin Aghamore, Joanne Bennette,
                Michelle Keller, Christine Mathosian
                and Tammy Sauter
Circulation: Iyatunde Abdullah, Cathy DeGregory and
              Annette Hellinger

Library of Congress Cataloging–in–Publication Data
available upon request

ISBN 1-56072-375-0
© 1997 Nova Science Publishers, Inc.
6080 Jericho Turnpike, Suite 207
Commack, New York 11725
Tele. 516-499-3103 Fax 516-499-3146
E-Mail: Novascience@earthlink.net

Printed in the United States of America

CONTENTS

A. Introduction

The term "inhibition" as applied to synaptic transmission refers to an active process by which excitatory transmission is prevented. At present it seems quite well understood that it is through the interaction between excitatory and inhibitory events that the integrative functioning of the nervous system is carried out (cf. Eccles, 1957, 1964; Katz, 1966).

Central inhibition was first demonstrated experimentally and described by Sechenov in 1863. Subsequently the phenomenon of somatic motor inhibition in vertebrates was further investigated and thoroughly analysed by Sherrington (1906) who arrived at a conclusion that somatic inhibition must be of central origin since there are no inhibitory nerves supplying the skeletal muscles in contrast to the innervation of visceral muscles; in addition it was inferred that the lowest (or most peripheral) level at which such inhibition can be initiated is the anterior horn cell or its cranial homologue (cf. Hebb, 1959).

Our knowledge of the basic mechanisms at the synapses of central neurons is largely derived from studies on mammalian spinal motoneurons (Frank and Fuortes, 1955; Eccles, 1957 1964; Fatt, 1957). As for the inhibitory effects, it has been clearly demonstrated that there are at least two types of central nervous system inhibition: postsynaptic and presynaptic. Postsynaptic inhibition is mediated by synapses placed directly on the membrane of the postsynaptic cell.

With presynaptic inhibition, the inhibitory synapses are located on the presynaptic endings of excitatory nerve fibers going to the postsynaptic cell. All evidence initially obtained in studies of spinal motoneurons (Brock, Coombs and Eccles, 1952; Coombs, Eccles and Fatt, 1955a,b; Eccles, 1957, 1964) strongly suggested that postsynaptic inhibition is associated with hyperpolarization of the neuron: the membrane potential change called inhibitory postsynaptic potential (IPSP). This IPSP is produced by an increase of membrane permeability to Cl^- or Cl^- and K^+ ions under the influence of the inhibitory synaptic transmitter. But soon after the discovery of IPSPs in motoneurons it was found that the postsynaptic influence could not account for the whole of the inhibitory effects of afferent nerve impulses on motoneurons (Frank and Fuortes, 1957; Eccles, 1961). As for another type of central nervous system inhibition called presynaptic inhibition, its mechanisms appeared to be quite different from that of postsynaptic inhibition. In this case the postsynaptic membrane is not affected in any detectable way; the inhibition is produced by a reduction of the amount of the synaptic transmitter released by a depolarization of presynaptic terminals of the excitatory fibers. This phenomenon closely resembles that demonstrated by Dudel and Kuffler for the crustacean neuro-muscular system (1961).

It seems rather natural to generalize these data on the central inhibition and to expand them to other levels of the central nervous system too, the neocortex included. At the same time some marked peculiarities of the inhibitory phenomenon established during investigation of higher brain structures (in particular of the neocortex) must be kept in mind. As reasonably suggested by Jung (1967), cortical inhibitory mechanisms appear to be predominantly postsynaptic, not presynaptic. Furthermore it is very important to mention (Jung, 1967) that every cerebral neuron thus far investigated with intracellular recording in the isocortex (Lux and Klee, 1962), in the allocortex

(Kandel and Spencer, 1961), and in the thalamus (Andersen and Eccles, 1962) exhibited hyperpolarizing membrane inhibition of long duration lasting nearly a hundred times longer than the ISPSs of motoneurons.

According to literature (Voronin, 1970; Takata and Ogata, 198Q; Takata, 1981, 1982; Satou, Mori, Tazawa and Takagi, 1982; Roy, Clercq and Steriade, 1984; Kehl and McLennan, 1985a,b; Avoli, 1986; Z. G. Kokaia, Kokaia, Labakhua and Okujava, 1986), postsynaptic inhibition of neurons in the brain in the form of hyperpolarization of a cellular membrane must be a rather complex, non-uniform process. A considerable part of this hyperpolarization seems to depend upon the increased permeability of the postsynaptic membrane to chloride ions (Krnjevic, Randic and Straughan, 1964, 1966a,b,c; Roy, Clercq and Steriade, 1984; Avoli, 1986; Z. G. Kokaia, Kokaia, Labakhua and Okujava, 1986; Serkov, 1986) as described for other levels of the nervous system (Coombs, Eccles and Fatt, 1955; Eccles, 1957, 1964). However, a substantially increased part of potassium permeability in this process attracts attention (Lancaster and Wheal, 1984; Roy, Clercq and Steriade, 1984; Avoli, 1986; Z. G. Kokaia, Kokaia, Labakhua and Okujava, 1987a,b, 1988). As far as various types of K^+ channels have been discovered in the membrane of a postsynaptic neuron (Hodgkin and Huxley, 1952; Meech and Standen, 1975; Meech, 1978; Newberry and Nicoll, 1984), the problem of the probable role of diverse K^+ channels in the generation of postsynaptic inhibitory responses in neocortical cells is of particular interest.

In addition to such a common mode of inhibition conditioned by increased membrane permeability and development of the conductance IPSPs under the influence of the chemical transmitter, the existence of another, less common mode of chemical inhibition, has also been described. It differs from the conductance IPSP by being directly dependent upon metabolism, upon the electrogenic pump, and not involving an increased conductance to ions (Nishi and Koketsu, 1967,

 V.M. Okujava

1968; Pinsker and Kandel, 1969; Spencer and Kandel, 1969). The likelihood of the occurrence of the nonconductance electrogenic pump IPSPs, operating in cortical neurons, remains still not fully investigated.

Relatively scarce are *in vivo* studies of the origin of some other hyperpolarizing events in neocortical neurons, such as afterhyperpolarizations following single or burst discharges of action potentials. Moreover, the association of such hyperpolarization potentials with synaptic inhibitory effects, in our view, has not attracted yet adequate attention.

And finally it must be mentioned that inhibition plays an essential role in initiation, development and cessation of seizure activity (Okujava, 1967, 1969, 1980, 1987; Prince and Wilder, 1967; Prince, 1968; Matsumoto, Ayala and Gumnit, 1969, Speckmann and Gutnick, 1992). Thus apparently further understanding of the intrinsic mechanisms of cortical inhibition will substantially contribute to elucidation of cellular basis of cortical epileptogenesis.

This book is devoted to the results of *in vivo* experimental studies by the author and his coworkers dealing with the ionic mechanisms and some pharmacological properties of inhibition in neocortical neurons. The details and methods of investigation can be found elsewhere (Okujava, 1969, 1987; Li, Okujava and Bak, 1971; Labakhua and Okujava, 1992).

B. Inhibitory Synaptic Action In The Neocortex

1. General Features

Intrinsic cellular and ionic mechanisms of cortical inhibition have been thoroughly investigated by a great number of authors in various areas of the cerebral cortex: motor (Phillips, 1956, 1959; Stefanis and Jasper, 1964a,b; Voronin, 1967) somatosensory (Andersson, 1965; Oscarsson, Rosen and Sulg, 1966; Innocenti and Manzoni, 1972; Storozhuk, 1974), visual (Li, 1959; Skrebitski and Voronin, 1966; Watanabe, Konishi and Creutzfeldt, 1966; Creutzfeldt and Ito, 1968; Creutzfeldt, Kuhnt and Benevento, 1974; Toyoma, Matsunami, Ohno and Tokashiki, 1974; Skrebitski and Sharonova, 1972), sensorimotor (Whitehorn and Towe, 1968; Voronin, 1970; Shuranova and Gvozdikova, 1971; M. G. Kokaia, Labakhua and Okujava, 1984), auditory (Serkov, 1975, 1977, 1984, 1986; Yanovski 1978; Ribaupierre, Goldstein and Komshian, 1972), associative (Dubner and Rutledge, 1965; Artemenko and Mamonets, 1972; Kazakov and Izmestiev, 1972; Kazakov, Izmestiev and Perkhurova, 1972; Mamonets, 1981; Yanovski, 1986), olfactory (Biedenbach and Stevens, 1969; Haberly, 1973; Galvan, Grafe and Bruggencate, 1982; Galvan, Franz and Constanti, 1985; Satou, Mori, Tazawa and Takagi, 1982) cortices. Similar studies were performed on slices of the sensorimotor cortex of rats (Avoli, 1986) and guinea pigs (Connors, Gutnick and

Prince, 1982; McCormick, Connors, Lighthall and Prince, (1985). These studies have revealed that inhibition of neuronal discharges, usually observed in extracellular recordings, are accompanied by hyperpolarization in intracellular recordings (Mountcastle and Powell, 1959; Powell and Mountcastle, 1959; Kondratyeva, 1967; Vassilevski, 1968; Storozhuk, 1974), which must be an active process. Such hyperpolarization has been considered to be an inhibitory postsynaptic potential (IPSP).

In neocortical neurons ISPSs can be evoked by peripheral sensory stimulation (Andersson, 1965, Voronin, 1967; Voronin and Skrebitski, 1967; Whitehorn and Towe, 1968; Voronin and Tanenholtz, 1969; Voronin and Ezrokhi, 1971; Ribaupierre, Goldstein and Komshian, 1972; Shaban, 1972; Serkov, 1975) or by means of electrical stimuli applied to various cerebral structures: pyramidal tracts (Phillips, 1959; Stefanis and Jasper, 1964a,b; Pollen and Lux, 1966; Humphrey, 1968; Renaud, Kelly and Provini, 1974; Raabe and Gumnit, 1975; Mzhavia, Okujava, 1980; Okujava, 1987; Labakhua and Okujava, 1992), cortical surface (Krnjevic, Randic and Straughan, 1966; Dreifuss, Kelly and Krnjevic, 1969; Kelly, Krnjevic, Morris and Yim, 1969; Shuranova and Gvozdikova, 1971; Fanarjian and Roitbak, 1974; Labakhua, Bekaya and Okujava, 1982; Okujava, Mzhavia and Goff, 1983; M. G. Kokaia, Labakhua and Okujava, 1984; Labakhua and Okujava, 1992), transcallosal pathways (Asanuma and Okamoto, 1959; Toyoma, Tokashiki and Matsunami, 1969; Nakamura, Naito, Kurosaki and Tamura, 1971; Storozhuk, 1974; Mamonets, 1981), thalamic relay nuclei (Li, 1963; Nacimiento, Lux and Creutzfeldt, 1964; Purpura and Shofer, 1964; Creutzfeldt, Lux and Watanabe, 1966; Creutzfeldt, Watanabe and Lux, 1966; Storozhuk, 1974; Serkov, 1977; Labakhua, Bekaya and Okujava, 1982; M. G. Kokaia, Labakhua and Okujava, 1984; Labakhua and Okujava, 1992) as well as thalamic nonspecific nuclei (Labakhua and Okujava, 1992).

In each of these cases the IPSPs are manifested as long-lasting hyperpolarization waves (Fig. 1) with suppression of both background and evoked spike discharges. It has been shown (Stefanis and Jasper, 1964; Okujava, Mzhavia and Goff, 1983) that during the IPSP, a marked drop of membrane resistance is observed, and neuronal excitability appears substantially decreased as measured by means of intracellularly injected depolarising current pulses (Fig. 2 and fig. 3).

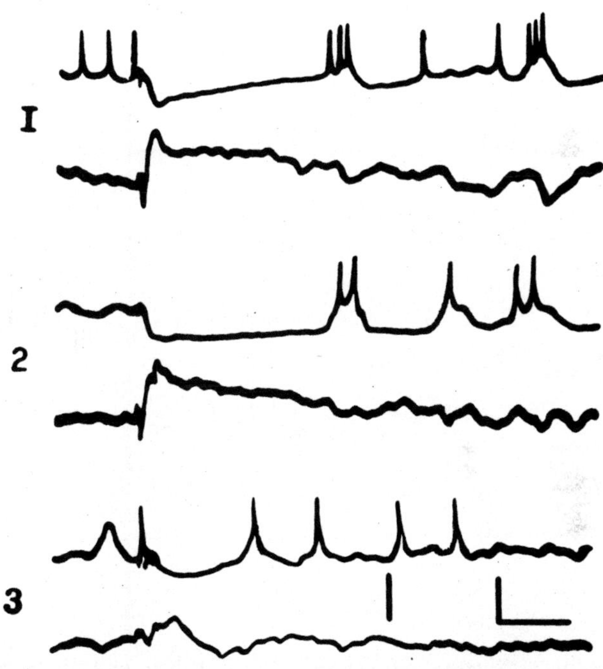

Figure 1. Intracellular neuronal responses (upper traces) and evoked cortical potentials (lower traces) elicited in the sensorimotor cortex by stimulation of various structures: direct cortical stimulation (1), nonspecific thalamic stimulation (2), and antidromic stimulation of medullary pyramids. Note long duration hyperpolarizing waves in intracellular records. Vertical bars represent 20 mV for microelectrode and 200 mV for gross electrode recordings; horizontal bar, 100 msec (Labakhua, Bekaia and Okujava, 1982).

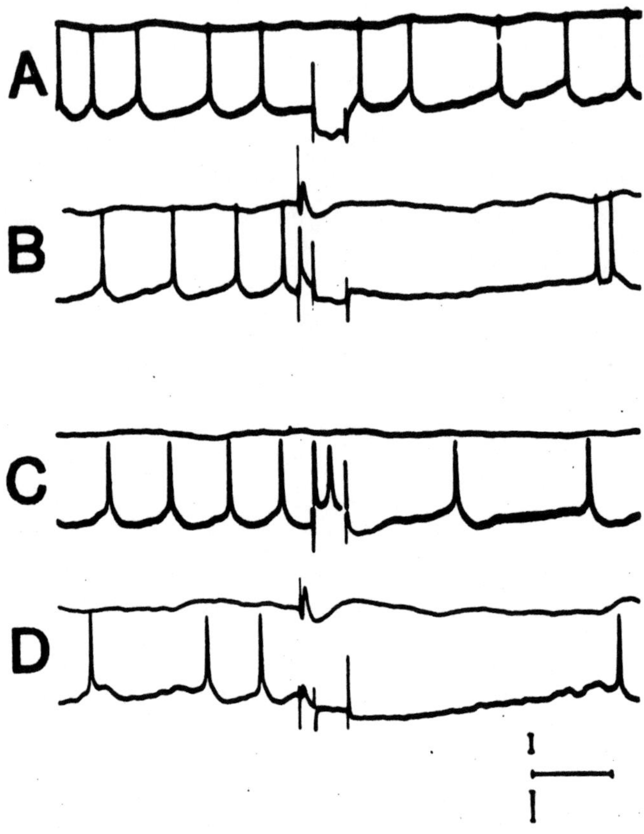

Figure 2. Testing of membrane resistance and neuronal excitability during hyperpolarization of a sensorimotor cortical cell evoked by direct cortical stimulation. Upper traces represent gross electrocorticographic recording and lower traces—intracellular recording. A. hyperpolarizing current pulse injected during hyperpolarization wave of the membrane potential. C. Depolarizing current pulse injected during background activity. D. Depolarizing current pulse injected during hyperpolarization wave. Vertical bars represent 100 mV for gross electrode and 10 mV for microelectrode recordings; horizontal bar, 100 msec (Okujava, Mzhavia and Goff, 1983).

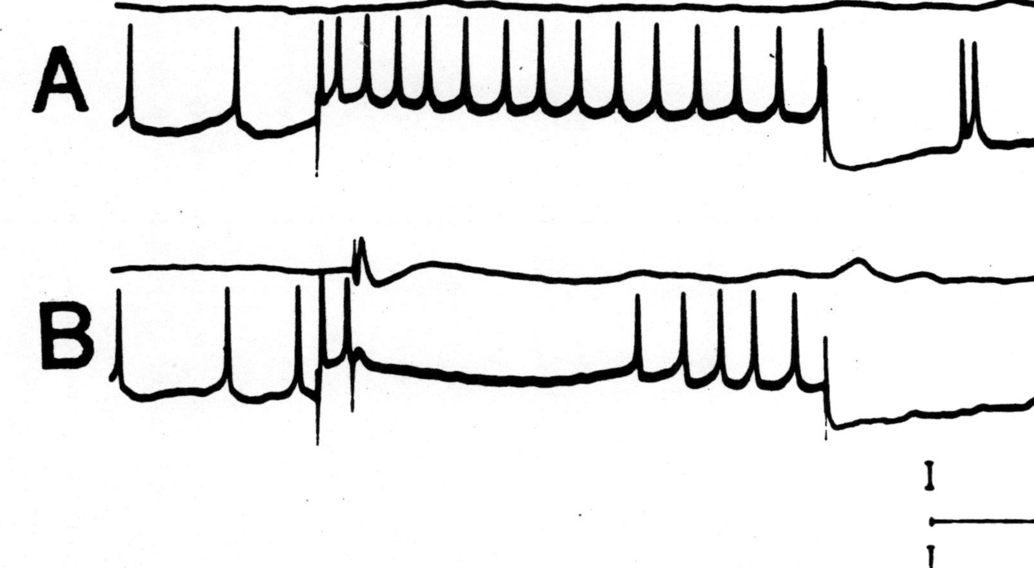

Figure 3. Neuronal reaction to intracellularly injected depolarizing current (A); Neuronal response to direct cortical stimulation during intracellular current injection (B). Upper traces monitor cortical surface activity; lower traces, intracellularly recorded neuronal activity. Vertical bars represent 100 mV for gross electrode and 10 mV for microelectrode records; horizontal bar, 100 msec (Okujava, Mzhavia and Goff, 1983).

The duration of the IPSPs in cortical neurons, although it may vary widely in various situations, appears to be extremely long as compared to the IPSPs of spinal motoneurons, where underlying mechanisms have been thoroughly investigated and understood (Eccles, 1957, 1964, 1966, 1969). Phillips (1961) recorded IPSPs of 500 msec duration in some cortical neurons in response to pyramidal tract stimulation. In similar experiments of Stefanis and Jasper (1964a,b) duration of such potentials equalled, on average, to 100-150 msec. In visual cortical neurons stimulation of optic nerves evoked IPSPs of 100-200 msec duration (Watanabe, Konishi and Creutzfeldt, 1966). Comparable results were obtained by Krnjevic, Randic and Straughan (1966a,b,c), Fanarjian and Roitbak (1974), Storozhuk (1974), Mamonets (1981), Serkov (1986) and others in various cortical regions during stimulation of relay nuclei or afferent pathways at various levels or during superficial cortical stimulation. Moreover in slices of the sensorimotor cortex of guinea pigs (Connors, Gutnick and Prince, 1982) and rats (Avoli, 1986) 150-2500 msec and 200-1000 msec IPSPs were recorded respectively. Extremely well pronounced and long-lasting potentials were revealed during neocortical paroxysmal epileptic activity (Okujava, 1967a,b, 1969, 1980, 1987).

It has been postulated (Creutzfeldt, Lux and Watanabe, 1966) that the long duration of cortical postsynaptic potentials favours temporal summation and integration of afferent impulses. At the same time it may be assumed that the electrophysiological peculiarity of cortical cells, namely their long time constant, is the main reason for the long duration of postsynaptic potentials, compared with that of motoneurons. In order to test the correctness of such an assumption the time constant of cortical neurons was measured (Li, Okujava and Bak, 1971). For this purpose hyperpolarizing currents of increasing intensity were applied through the intracellular electrode and the increase of membrane potential in response to the current was recorded. The time constant was measured as the time required to reach 0.632, of the final value of the resulting amplitude, as shown in Fig. 4. The results of such determination gave the mean value of 5.69 ± 1.34 msec. Somewhat higher, but quite comparable values were obtained by other authors: Creutzfeldt, Lux and Watanabe, (1966)—8.5 ± 2.2 msec, Lux and Pollen (1966)—8.4 ± 1.4 msec, Takahashi (1965)—6.7 ± 2.2 msec and 7.6 ± 2.3 msec in slow and fast groups of pyramidal tract (PT) cells respectively. These findings indicate that the extremely long duration of ISPSs, especially that of the descending phases, cannot be explained merely by passive properties of cortical neuronal membranes and it must be conditioned by some active physiological process. Because in the neocortex, in contradistinction to the spinal cord, few cells can be found that discharge a train of impulses in accordance with IPSPs of neighbouring neurons (Krnjevic, Randic and Straughan, 1966b; Stefanis, 1969), as do Renshow cells in the spinal cord, it has been concluded (Eccles, 1966; Creutzfeldt, Lux and Watanabe, 1966; Krnjevic, Randic and Straughan, 1966a,b,c; Pollen and Lux, 1966; Serkov, 1986) that the long duration of IPSPs in cortical neurons at least partially must be the result of persisting action of an inhibitory transmitter in the synaptic cleft.

The amplitude of various IPSPs in neocortical neurons may vary in a wide range. In the somatic cortex, in response to single stimuli applied to the thalamic ventral posterior lateral nucleus (VPL), IPSPs of 7-10 mV were recorded (Storozhok, 1974), which increased up to 25 mV during neuronal depolarization as a result of the injury by the microelectrode (Voronin, 1967). During the pyramidal tract stimulation the amplitude of IPSPs in pyramidal neurons varied between 2 and 15 mV (Stefanis and Jasper, 1964a). Somewhat similar magnitude (3-13 mV) was observed during cortical surface stimulation (Fanarjian and

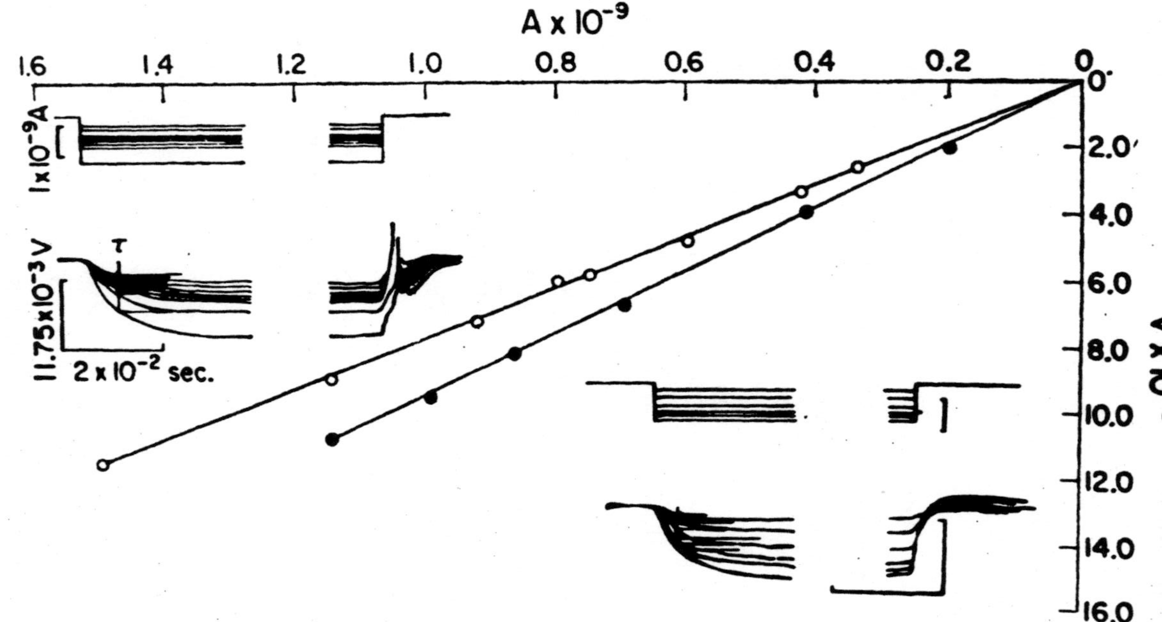

Figure 4. Current-voltage plot. Inserted tracings in left upper corner were obtained from superimposed recordings of responses to passing current of increasing intensity. The upper traces show the current through the micropipette and the lower traces show the responses of the cell. The current-voltage plot for this neuron is shown as open circles. Similarly for another neuron, in the right lower corner, as solid circles. (() Represents a time constant (Li, Okujava and Bak, 1971).

Roitbak, 1974). In neurons of the auditory cortex (Ribaupierre, Goldstein and Komshian, 1972) IPSPs of about 10 mV (more frequently 3-8mV) were evoked; in the visual cortex (Watanabe, Konishi and Creutzfeldt, 1966; Skrebitski, Sharonova, 1972) 4-10 mV potentials were recorded and in the associative cortex (Mamonets, 1981) in response to strong stimuli they attained 20 mV. The amplitude of the IPSPs in the sensorimotor cortical slices of guinea pigs averaged 3.0 ± 1.2 mV (Connors, Gutnick and Prince, 1982).

The amplitude of the IPSP depends to a considerable extent upon the magnitude of the membrane potential and upon the strength of stimulation. According to Krnjevic, Randic and Straughan (1966b) cortical stimulation of 30 V evoked in neurons only excitatory postsynaptic potentials (EPSPs). When the stimulation intensity was enhanced up to 50 V the EPSP was succeeded by a pronounced IPSP. At the stimulation intensity of 70 V the amplitude of the IPSP increased and attained 10 mV.

The dependence of the IPSP amplitude upon the intensity of stimulation was revealed in PT neurons (Stefanis and Jasper, 1964a). If the amplitude of the IPSP at 4 V of pyramidal tract stimulation amounted to 2.5 mV, at 10 V of stimulation it reached 5 mV.

It must be mentioned that the increase of the stimulation intensity simultaneously with augmentation of the amplitude of the IPSP produces its prolongation. This has been convincingly shown by Serkov (1986) for auditory cortical neurons during stimulation of geniculocortical fibers. The threshold stimulation evoked 1-1.5 mV IPSPs of 5-10 msec duration. Gradual enhancement of the strength of stimulation led to an increase of both the amplitude and the duration of the IPSP so that the quadrupled strength of stimulation elicited the 6.6 mV IPSP of 60 msec duration. Similar changes of the IPSP length in connection with the stimulus strength was also described by other authors (Krnjevic, Randic and Straughan 1966b; Dreifuss, Kelly and Krnjevic, 1969).

As already mentioned above, the amplitude of the cortical IPSP depends not only upon the strength of stimulation but also upon the membrane potential value at the moment of recording, the latter varying on average between 40 mV and 70 mV (Phillips, 1956, 1959; Stefanis and Jasper, 1964a; Krnjevic, Randic and Straughan, 1966b; Lux and Pollen, 1966; Watanabe, Konishi and Creutzfeldt, 1966; Voronin, 1967; Creutzfeldt and Ito, 1968; Li, Okujava and Bak, 1971; Serkov, 1986).

Stefanis and Jasper (1964a) noticed close dependence of the IPSP amplitude in the PT neurons upon the spontaneous or artificial changes of the membrane potential. So its amplitude was 2.5 mV when the membrane potential equalled to -65 mV and it increased up to 5 mV during depolarization to -58 mV. Unfortunately in this study the influence of membrane hyperpolarization was not tested, as it had been done earlier on spinal motoneurons (Coombs, Eccles and Fatt, 1955) showing that hyperpolarization of the neuronal membrane over -82 mV leads to the reversal of the IPSP.

Similar events were studied later in neocortical neurons too. Purpura and Shofer (1964) when investigating responses of motor cortical neurons to thalamic stimulation revealed reversal of their IPSPs during polarization of the cortex with direct current. It must be mentioned that under the influence of hyperpolarizing currents reversal of the initial part of the IPSP was mainly observed (Purpura, 1973). Reversal of the IPSPs in these neurons was also established later in cases of membrane hyperpolarization by means of current injection through the intracellular microelectrode (Stefanis, 1969). The IPSPs markedly increased at a background of depolarization and they decreased and became reversed under the influence of hyperpolarization (Voronin, 1967, 1970, Labakhua and Okujava, 1992).

A similar behaviour was described for the IPSPs in visual cortical neurons evoked by optic nerve stimulation (Watanabe, Konishi and Creutzfeldt, 1966) and for the IPSPs in auditory cortical neurons evoked by acoustic stimuli (Ribaupierre, Goldstein and Komshian, 1972). In the association cortex, too, the amplitude of the IPSP appeared to depend upon the level of the membrane potential (Yanovski, 1986). A same event was observed in neurons of the pyriform cortex of rabbits using intracellular injection of the polarizing current (Satou, Mori, Tazawa and Takagi, 1982). Similar data were obtained in neurons in the slices of the sensorimotor cortex of guinea pigs (Connors, Gutnick and Prince, 1982) and rats (Avoli, 1986).

In Fig. 5 the influence of the artificial polarization of the membrane upon the IPSP of the PT neuron is presented. At a resting potential level single electric shock to the thalamic VPL nucleus evokes the EPSP of 3.5 mV followed by the 9 mV IPSP of 400 msec duration. During injection of the hyperpolarizing current through the intracellular microelectrode the gradual decrease of the IPSP amplitude and increase of the EPSP amplitude is observed. At a current strength of -1.5 nA the IPSP becomes totally depressed while the EPSP amplitude attains 12 mV. Artificial depolarization of the membrane produces enhancement of the IPSP. At a current strength of ±1.8 nA it attains 20 mV.

These data indicate that different IPSP amplitudes in various cortical neurons in response to a certain stimulus can be conditioned by different values of the membrane potential in these neurons. According to Serkov (1986) the difference in the IPSP amplitudes in neurons with identical and normal membrane potentials in response to identical stimuli can be attributed to the different numbers of inhibitory synapses activated upon the neurons.

V.M. Okujava

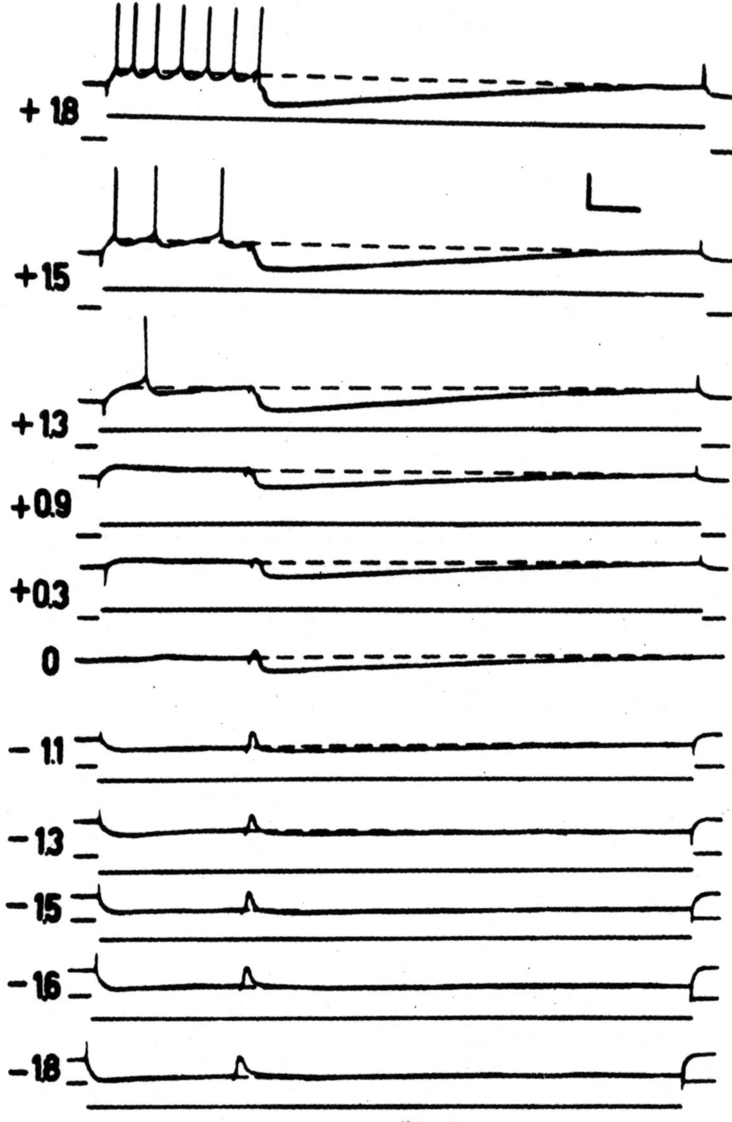

Figure 5. The influence of artificial polarization of a pyramidal neuron upon the IPSP. Numbers on the left indicate the magnitude of the polarization current in nA (injection signalled by the lower trace). Vertical bar represents 25 mV; horizontal bar, 50 msec, (According to studies performed with Z. G. Kokaia, 1987).

2. THE COMPOSITE NATURE OF THE IPSPs

According to various experimental data the long-lasting cortical IPSP appears not to be a phenomenon of homegenous nature along its full length. It seems to consist at least of two components of differing origin: the early and the late ones. Analogous structure of neuronal inhibitory potentials has been proposed for various levels of the central nervous system: the spinal cord and the brain stem (Goldberg and Nakamura, 1968; Kidokoro, Kubota, Shuto and Sumino, 1968; Cook, Duncan and Cangiano 1971; Takata and Ogata, 1980; Takata, 1981), the hippocampus (Alger and Nicoll, 1982; Alger, 1984; Fournier and Crepel, 1984; Newberry and Nicoll, 1984; Kehl and McLennan, 1985a,b; Blaxter, Carlen, Davies and Kujtan, 1986), the thalamus (Roy, Clercq and Steriade, 1984), the pyriform cortex (Satou, Mori, Tazawa and Takagi, 1982). A wealth of experimental proofs has accumulated in favour of two-componental composition of the IPSPs in the neocortical neurons (Stefanis, 1969; Voronin, 1970; Serkov, 1984; M. G. Kokaia, Labakhua and Okujava). In that respect a number of pharmacological and electrophysiological investigations deserve attention.

a) The influence of strychnine upon cortical inhibition. The blocking action of strychnine upon postsynaptic inhibition in the central nervous system has been particularly well investigated for the spinal cord (Coombs, Eccles and Fatt, 1955a,b; Eccles, 1957, 1964; Fuortes and Nelson, 1963; Curtis, Hosli and Johnston, 1968a,b; Curtis, Duggan and Johnston, 1969, 1971; Phillis, 1970; Takata and Ogata, 1980; Bagust, Green and Kerkut, 1981; Schwindt and Crill, 1981). In addition to its antiinhibitory effect during intravenous injections, Curtis (1962) has been able to demonstrate its depressant action on postsynaptic inhibition following electrophoretic administration.

The results of the experimental studies of the influence of strychnine upon cortical inhibition appear to be more controversial. According to a number of investigations with both extracellular (Suzuki and Tukahara, 1963) and intracellular recordings (Pollen and Ajmone Marsan, 1965; Stefanis and Jasper, 1965; Pollen and Lux, 1966; Sawa, Maruyama, Kaji and Nakamura, 1966; Storozhuk, 1974; Avoli, Barra, Brancati, Deodati and Vagnozzi, 1976) strychnine exerts depressive action upon inhibitory neuronal events—spontaneous and evoked. The blockade of the IPSP in cortical neurons was also revealed during extracellular iontophoretic application of strychnine simultaneously with intracellular recording (Okujava, 1975).

However, contrary to these data, other observations exist indicating that strychnine does not influence cortical inhibitory processes or it exerts selective action only upon some forms of inhibitory reactions.

Thus, after topical application of strychnine to the cerebral cortex, in addition to neurons with depressed hyperpolarizing electrogenesis, Li (1959) recorded neurons with well pronounced hyperpolarizing potentials arising in correlation with electrocorticographic (ECoG) paroxysmal discharges.

In experiments with extracellular microelectrodes Crawford, Curtis, Voorhoeve and Wilson (1963) following intravenous strychnine injection could not reveal any changes of inhibitory responses of Betz cells to the pyramidal tract or direct cortical stimulation.

According to Brooks and Asanuma (1965a) intravenous injections of subconvulsive doses of strychnine reduced recurrent inhibition slightly and reversibly in the pericruciate cortex.

In studies performed by Krnjevic, Randic and Straughan (1966c) neither intravenous injection of 0.1-0.6 mg/kg strychnine nor its application as a 1% solution to the cortical surface and nor even its iontophoretic administration through the micropipette to the recorded neuron produced any noticeable reduction of an inhibitory response evoked by direct cortical stimulation.

Moreover Phillis and York (1967a,b, 1968a,b) and later Jordan and Phillis (1972) using iontophoretic application of strychnine to the extracellularly recorded neuron, found the blockade of synaptic inhibition in some cases of stimulation and its absence in other ones.

The following experimental findings, obtained in the author's laboratories (M. G. Kokaia, Labakhua and Okujava, 1981, 1982; Mzhavia, 1982; M. G. Kokaia, 1983), may serve as an explanation of the above seemingly contradictory results.

After topical application of strychnine to the cortical surface in neurons of deep cortical layers, simultaneously with ECoG seizure discharges, very well pronounced hyperpolarization potentials (presumably IPSPs) are recorded, succeeding EPSPs or depolarizing paroxysmal shifts (PDSs). As shown in Fig. 6 A, B, the neuronal IPSPs coincide with spontaneous ECoG discharges. The same situation is revealed during cortical stimulation with single shocks (Fig. 6 B, D). But at the same time it must be emphasized that such results were obtained in neurons of deep cortical layers (1.5 mm and deeper from the cortical surface) while in other, more superficially located neurons noticeable reduction of the IPSPs could be revealed. In Fig. 7 the IPSPs, evoked both by thalamic *n. ventralis lateralis* (VL) and direct cortical stimulation, appear reduced (mostly their initial parts) for some 33% following topical application of strychnine solution. Still in another neuron (Fig. 8.3) the early part of the IPSP becomes substituted by the depolatization wave, later (Fig. 8.24) transformed into the PDS. A twofold reduction of the stimulation intensity leads to restoration of the depolarization EPSP in the initial part of the cellular response (Fig.8.52). Thus on the basis of these findings, the blocking influence of strychnine upon the cortical IPSPs (at least their initial parts) can be assumed. In all probability strychnine mostly exerts its action in the site

                          V.M. Okujava

of its immediate application—superficial layers—causing there paroxysmal seizure discharges. In more distant regions hyperpolarizing events appear more potent creating thus the inhibitory surround (Prince and Wilder, 1967; Prince, 1968; Dichter and Spencer, 1969; Okujava, 1980, 1987) (cf. Chapter 4). Hyperpolarizations of long duration in nerve cells of deeper cortical layers must be considered as an inhibition which is directed from superficial laminae to the deeper ones, the phenomenon described as "vertical inhibition" (Elger and Speckmann, 1983a,b, 1984, 1987; Speckmann, 1986; Pockberger and Speckmann, 1989) Thus from such findings it can be derived that the experimental epileptic focus (strychnine focus in this particular case) is enveloped hemispherically by an inhibition zone. It can be assumed that this is the basis of spatial restriction of seizure activity (Traub, 1983).

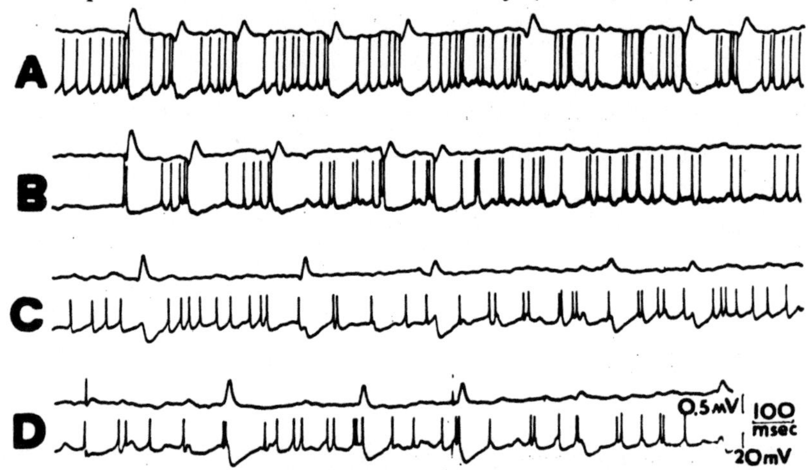

Figure 6. Effect of topical application of strychnine to the cortical surface in sensorimotor region. Intracellular neuronal activity in deep cortical layers (lower trace) and ECoG (top trace) recorded several minutes following application of 1% strychnine solution. In A, B, and C, D two various neurons are explored. A, C. Intracellular records during spontaneous intermittent ECoG paroxysmal discharges. B, D. Intracellular records during cortical stimulation with single electrical shocks. Note EPSPs followed by prolonged hyperpolarization waves in accordance with both spontaneous and evoked ECoG discharges (According to studies performed with Mzhavia, 1982).

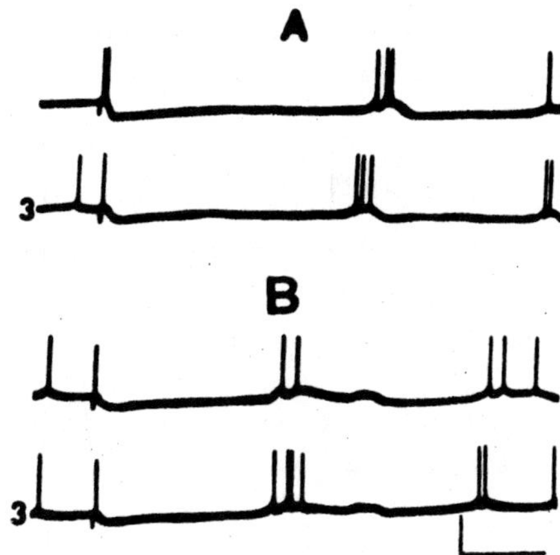

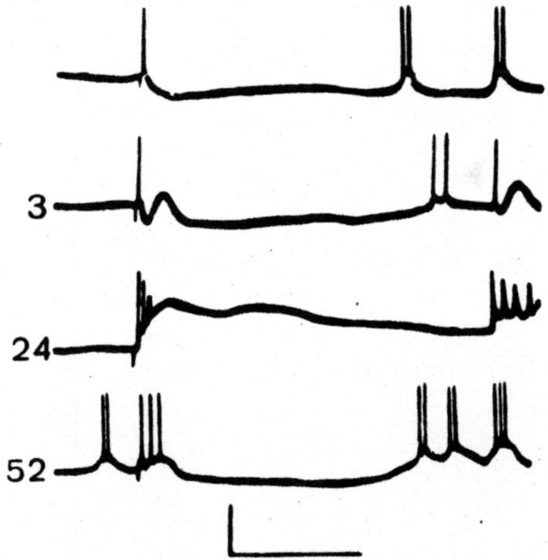

Figure 7. The influence of topical application of strychnine to the cortical surface upon the IPSPs evoked by thalamic VL (A) and direct cortical (B) stimulations in pyramidal tract neurons. Numbers on the left indicate time in min after application of strychnine. Vertical bar, 60 mv; horizontal bar, 100 msec (M. G. Kokaia, Labakhua and Okujava, 1982).

Figure 8. The influence of superficial cortical application of strychnine upon the PT cell response to VL stimulation. Numbers of the left indicate time in min after application of strychnine solution. Vertical bar, 5 mV; horizontal bar, 100 msec (M. G. Kokaia, Labakhua and Okujava, 1982).

As a matter of fact the IPSPs in deep cortical neurons appear markedly reduced when, instead of superficial application, strychnine solution is injected into the cortex. In Fig. 9 such a case is presented with spontaneous seizure discharge (A) and with responses to direct cortical (B) and to pyramidal tract (C) stimulations.

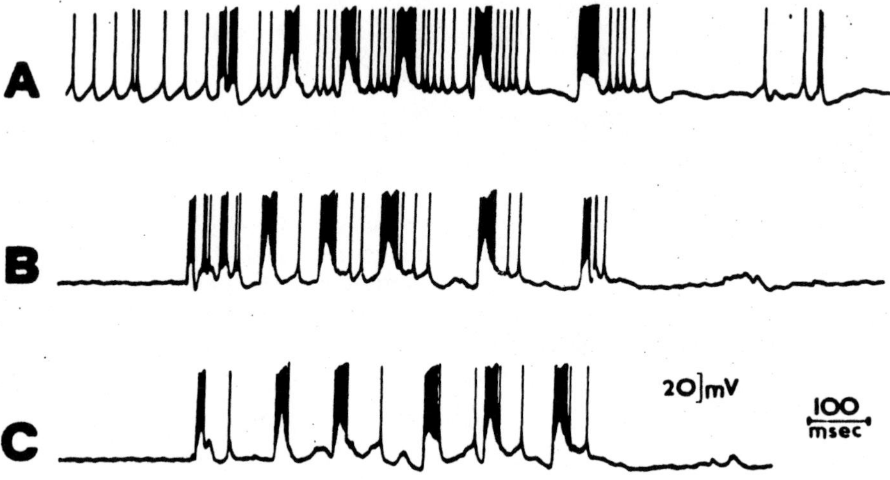

Figure 9. The influence of intracortical injection of 1% strychnine solution upon the PT neuronal activity in the cortical depth. A. Spontaneous paroxymal activity. B. Neuronal response to pyramidal stimulation at a bulbar level (According to studies performed with Mzhavia, 1982).

More accurately the action of strychnine upon synaptic events can be tested in experiments with intracellular recordings of the neurons and extracellular iontophoretic application of the substance. This can be effectively carried out by means of a special double-barrelled micropipette with an intertip distance of approximately 50-60 μm along its longitudinal axis. It appeared that the IPSPs, whatever their triggering stimuli are, undergo quite identical changes under the influence of iontophoretically injected strychnine. In all these cases strychnine leads to the reduction of the IPSP (Fig. 10), but it must be mentioned that the reduction concerns only the initial segment of the IPSP while the later part remains fully unchanged in its amplitude and duration. Moreover, the very initial phase of the IPSP may become replaced be depolarization potential, which frequently reaches the level for spike generation. In Fig. 11 it is seen that iontophoretic application of strychnine produces reduction of the initial segments of the IPSPs evoked by VPL and direct cortical stimulations and their replacement by depolarizations leading to short bursts of action potentials while the later parts of the IPSPs remain unchanged, inhibiting efficiently spike activity of the neuron. The blockade of the IPSPs lasts long enough depending on the strength and duration of the injecting current. In Fig. 12 partial restoration of the early component of the IPSP is observed 3 min later after cessation of the iontophoretic current.

Substitution of the initial segment of the IPSP by depolarization during strychninization of the cortex has been noticed in a number of other investigations too (Pollen and Ajmone Marsan, 1965; Pollen and Lux, 1966; Stefanis and Jasper, 1965). It was assumed (Pollen and Ajmone Marsan, 1965) that under the influence of strychnine ionselective channels, usually activated by inhibitory transmitter, become damaged, lose their selectivity and start to pass the greater

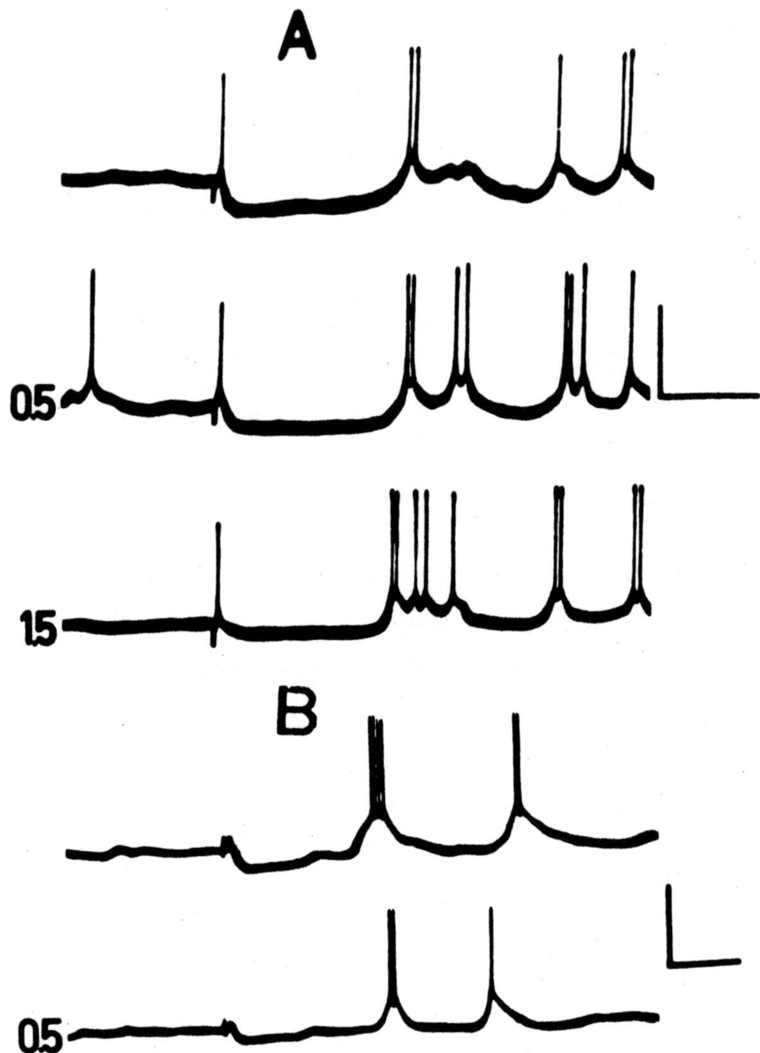

Figure 10. The influence of iontophoretic application of strychnine upon ISPSs in response to cortical neurons (A and B) to the thalamic VL stimulation. In A iontophoresis with 50 nA current lasted for 30 sec and in B for 1 min. Numbers on the left indicate time in min after the end of iontophoresis. Calibration: 50 mV; 100 msec (M. G. Kokaia, Labakhua and Okujava, 1982).

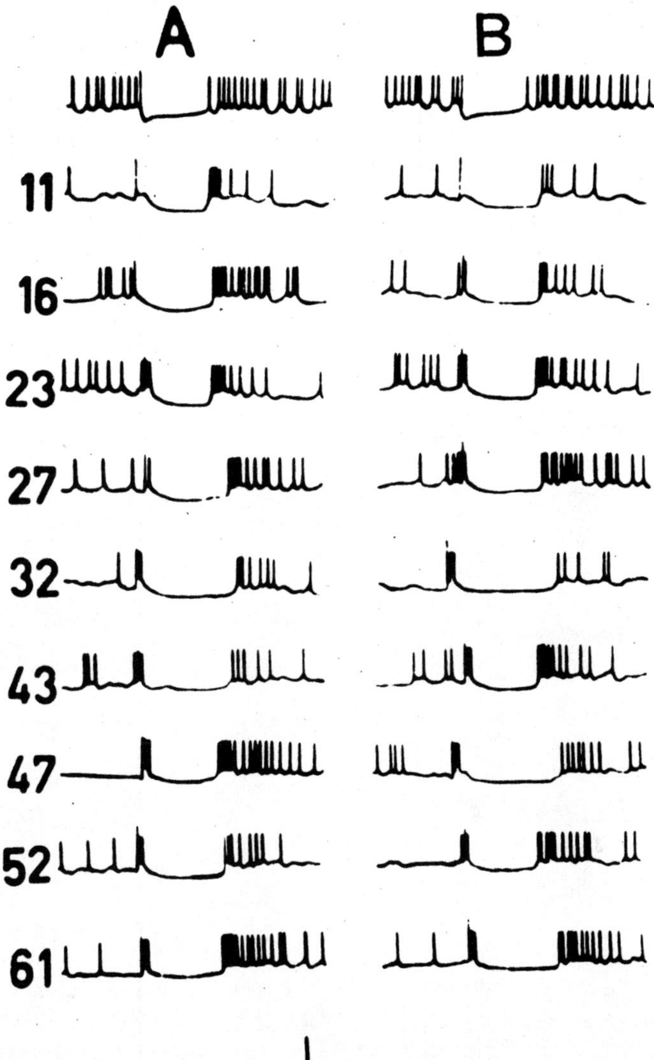

Figure 11. The influence of iontophoretic application of strychnine upon the responses of the PT neuron to the thalamic VL (A) and to the cortical surface stimulations (B). After recording of background responses iontophoresis was performed with 50 nA current for 2 min. Later it was repeated on the 20th min with 100 nA for 2 min and on the 48th min for 3 min. Numbers on the left indicate time in min after the end of the first iontophoretic session. Vertical bar, 20 Mv; horizontal bar, 100 msec (M. G. Kokaia, Labakhua and Okujava, 1981).

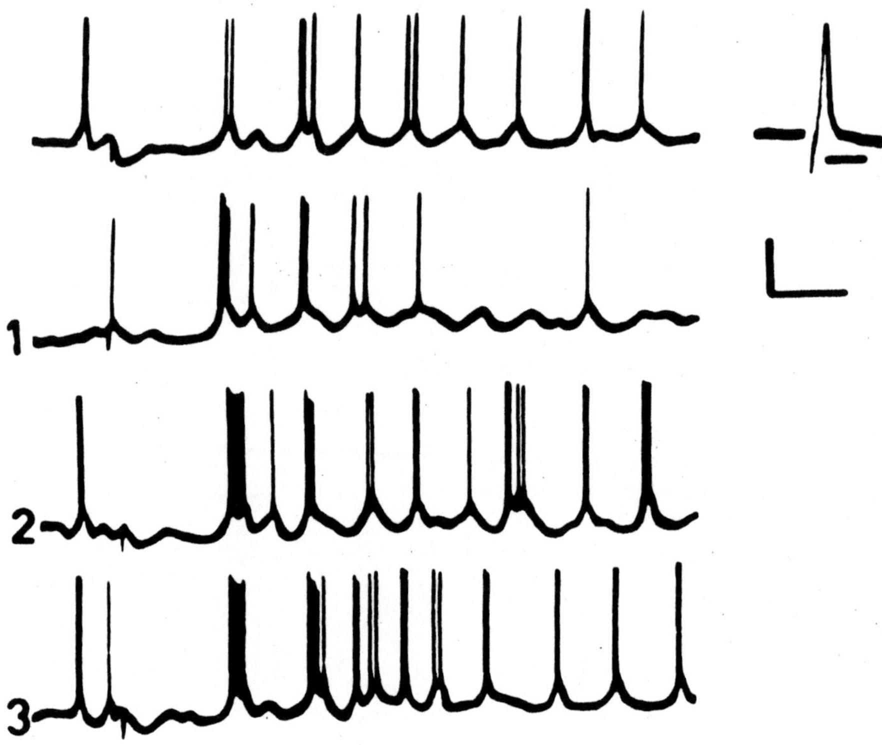

Figure 12. The influence of iontophoretic strychnine application upon the responses of a PT neuron to VL stimulation. Upper right insert shows an antidromic spike with the time scale of 3 msec. After recording the background response iontophoretic injection was performed with 200 nA current for 2 min. Numbers on the left indicate time in min after the end of iontophoresis. Calibration: 50 mV; 100 msec (According to studies performed with M. G. Kokaia, 1983).

hydrated sodium ions thus producing the transformation of the central inhibition into the central excitation. However this assumption seems rather improbable since such transformation is observed only with regard to the early component of the IPSP. More acceptable seems the supposition (Chang, 1953; Curtis, Duggan and Johnston 1971; M. G. Kokaia, Labakhua and Okujava, 1982, 1984; Levi, Bernardi, Cherubini, Gallo, Marciani and Stanzione, 1982; Labakhua and Okujava, 1992) that strychnine as a competitive blocker of inhibitory synapses assists in unmasking of EPSPs coexistent with IPSPs and usually concealed by the latter. All the more so as the IPSP, especially its initial part, is practically never evoked separately (Krnjevic, 1974a,b); it constitutes an EPSP-IPSP mixture with predominance of the latter, because inhibitory synapses are localized upon the cellular somata (Purpura, 1973), such strategic location (Eccles, 1969) enabling them to shunt excitatory currents more efficiently.

The knowledge of the nature of neurotransmitters acting at inhibitory synapses may considerably assist in understanding the above-described effects of strychnine. Solid evidence has by now been accumulated to show that glycine acts as a synaptic transmitter of inhibitory synapses at certain sites and that strychnine is a selective antagonist of glycine-induced postsynaptic inhibition (Curtis, Hösli and Johnston, 1968a,b, 1969, 1971; Eccles, 1969; Larson, 1969; Biscoe, Duggan and Lodge, 1972; Krnjevic, 1974a; Cuningham and Miller, 1980; Logan, Lovick, West and Wolsteniroft, 1980). The other transmitter substance with potent inhibitory action at central nervous system synapses is gamma-amino butyric acid (GABA) (Krnjevic and Phillis, 1963; Weinstein, Roberts and Kakefuda, 1963; Crawford and Curtis, 1964; Mangan and Whittaker, 1966; Salmoiraghi and Stefanis, 1967; Dreifuss, Kelly and Krnjevic, 1969; Iversen, Mitchell and Srinivasan, 1971; Hammerstad and Cutler, 1972; Roberts, Chase and Tower, 1976; Waterhouse, Moises and Woodward, 1980). Despite the fact that a considerable number of compounds has been described to date as transmitters of synaptic inhibition, glycine and GABA remain two main substances mediating inhibition in the synapses of the mammalian central nervous system. It is generally resumed that GABA mediates inhibition in the brain, whereas glycine is restricted to the brain stem and the spinal cord. In other words the neural axis is imagined to be divided into a rostral part in which GABA seems to be the only transmitter and a caudal part occupied by glycine receptors starting at the midbrain level (Eccles, Fatt and Koketsu, 1954; Biscoe and Curtis, 1967; Krnjevic and Schwartz, 1967; Curtis, 1968; Werman and Aprison, 1968; Curtis, Felix and McLellan, 1970). Their actions can be blocked by different antagonists: strychnine is the competitive blocker for glycine, blocking glycine receptors at nanomolar concentration, while GABA can be blocked by a variety of drugs in the micro and milimolar range.

On the basis of such a distribution of inhibitory transmitters in the central nervous system the conclusion can be drawn that strychnine is a convulsant mainly in the spinal cord only. However, as already discussed, strychnine for many decades was used as a very efficient drug to produce convulsive phenomena in the cerebral cortex. Moreover the experimental findings already presented in this chapter clearly show the sizeable modification of the IPSPs in cortical neurons under the influence of strychnine. In this respect, first of all, it must be stressed that the oversimplified map for distribution of glycine and GABA respectively in the spinal cord and in the brain does not correspond fully to the facts. Recent immunocytochemical and molecular biological studies (Betz, 1991) and microphysiological investigations (Dichter, 1980; Shirasaki, Klee, Nakaye and Akaike, 1991) have demonstrated a widespread distribution of glycine receptors throughout the central nervous system, including the olfactory bulb (Trombley and Shepherd, 1994). Even earlier the evidence was available (Kelly and Krnjevic,

1969; Johnson, Roberts and Straughan, 1970; Biscoe, Duggan and Lodge, 1972) in favor of inhibitory action of glycine upon cortical neurons. Furthermore, some rather convincing data exist indicating that in 15%-20% of cortical neurons in rats glycine acts as an inhibitory transmitter (Bernard), Cherubini, Marciani and Stanzione, 1979; Marciani, Stanzione, Cherubini and Bernardi, 1980; Levi, Bernardi, Cherubini, Gallo, Marciani and Stanzione, 1982). Besides, an overlapping distribution of GABA and glycine receptors in various areas of the brain has been described (Huck and Lux, 1987; Triller, Cluzeaud and Korn, 1987; Krishtal, Osipchuk and Vrublevsky, 1988; Cohen, Fain and Fain, 1989; Ito and Cherubini, 1990) and both transmitters have been shown to display the same anion selectivity (Barker and McBurney, 1979; Hamill, Borman and Sakmann, 1983). Because of these data the assumption seems acceptable that the early strychnine-sensitive component of the IPSP in cortical neurons must be conditioned by the activity of glycine-mediated inhibitory synapses (M. G. Kokaia, Labakhua and Okujava, 1984). This cannot be the only statement because selectivity of strychnine for glycine-sensitive receptors does not seem entirely specific (Phillis, 1970); it rather depends upon the concentration of the drug (Davidoff, Aprison and Werman, 1969; Curtis, Duggan and Johnston, 1971; Biscoe, Duggan and Lodge, 1972; Krnjevic, 1974a).

The latest concept of the GABA-mediated inhibition recognizes GABA as the neurotransmitter for two types of postsynaptic inhibition in supraspinal regions of the mammalian central nervous system (cf. Macdonald and Meldrum, 1995): rapid inhibition mediated primarily by GABA acting on GABA-A receptors and additional postsynaptic inhibition with a slower time course mediated by GABA acting at GABA-B receptors. It must be noted that GABA-A receptors regulate gating of the chloride ion channels while GABA-B receptors are coupled to potassium channels. Rather similar features of GABA-B receptors have been revealed in the spinal cord as well (Curtis and Lacey, 1994).

As for the influence of strychnine upon the GABA-mediated postsynaptic inhibition, it has been shown that despite the difference in sensitivity of glycine and GABA receptors to strychnine (nanomolar vs. milimolar), strychnine is still a much more effective blocker of the GABA-A receptors than other established GABA blockers (Klee, Shirasaki, Nakaye and Akaike, 1992). This may serve as a good basis for explaining the effects of strychnine upon the cortical neuronal IPSPs presented in this chapter.

Finally, without entering into detailed discussion, it may be concluded that at all events the results of the described experiments with strychnine action demonstrate the composite nature of the IPSPs in the neocortex, each one of them consisting of the early strychnine-sensitive and the late strychnine-resistant components partially overlapping each other.

b) Membrane resistance measurements in cortical neurons during IPSPs. Just as expected the membrane resistance of cortical neurons undergoes substantial reduction during the IPSP (Fig. 2). Similar reduction has been described earlier in a number of publications (Pollen and Lux, 1966; Krnjevic and Schwartz, 1967; Biedenbach and Stevens, 1969; Dreifuss, Kelly and Krnjevic, 1969; Krnjevic, 1974a; M. G. Kokaia, Labakhua and Okujava, 1981, 1982; Labakhua and Okujava, 1992). At the same time it must be noted that the membrane resistance drop is not uniform within the limits of the entire duration of the IPSP. Thus in neurons of the *in vitro* neocortical slices of rats Avoli (1986) observed a drastic fall of the input resistance in the course of the early phase of the IPSP, while the drop was much smaller within the limits of the late component. Similar findings were obtained by Satou, Mori, Tazawa and Takagi (1982) in neurons of the pyriform cortex of rabbits.

Some analogous results are worth mentioning here. As seen in Fig.

13.1, a single electric shock applied to the thalamic VPL evokes the EPSP-IPSP complex in the PT neuron of the sensorimotor cortex. The measurement of the input resistance reveals its marked drop during the early component of the IPSP with its full restoration during the late phase; the resistance reaches its background value considerably earlier before the termination of the IPSP (Fig. 13.2).

In Fig. 14 the average value of the input resistance measurements during the early and late components of the IPSPs in 10 PT neurons is graphically presented. The restoration of the input resistance is clearly evident during the late component of the IPSPs. This fact might be explained merely by the degree of the membrane permeability changes in parallel with the amplitude of the IPSP (Watanabe, Konishi and Creutzfeldt, 1966), if the resistance changes directly followed the amplitude. But as shown above, the resistance restoration takes place earlier than the IPSP is ended.

Thus the membrane resistance measurements reveal marked difference between the early and the late components of the IPSPs. The reason of such disparity between two components may be the difference in the sites of origin of these components in the neuron. As a matter of fact input resistance shifts measured by means of the intrasomatically located microelectrodes primarily indicate the permeability alterations of the somatic membrane while participation of the more remote dendritic regions must be negligible. Therefore it may be suggested that the early component is conditioned by the activity of axo-somatic inhibitory synapses, while the late component is evoked by the action of axodendritic synapses (perhaps located even on proximal dendrites) (Voronin, 1970; Purpura, 1973; Storozhuk, 1974; Satou, Mori, Tazawa and Takagi, 1982; M. G. Kokaia, Labakhua, Okujava, 1984; Avoli, 1986; Serkov, 1986). On the other hand the origin of the late component can be alternatively explained as a less common mode of inhibition, namely as a nonconductance electrogenic pump IPSP, which

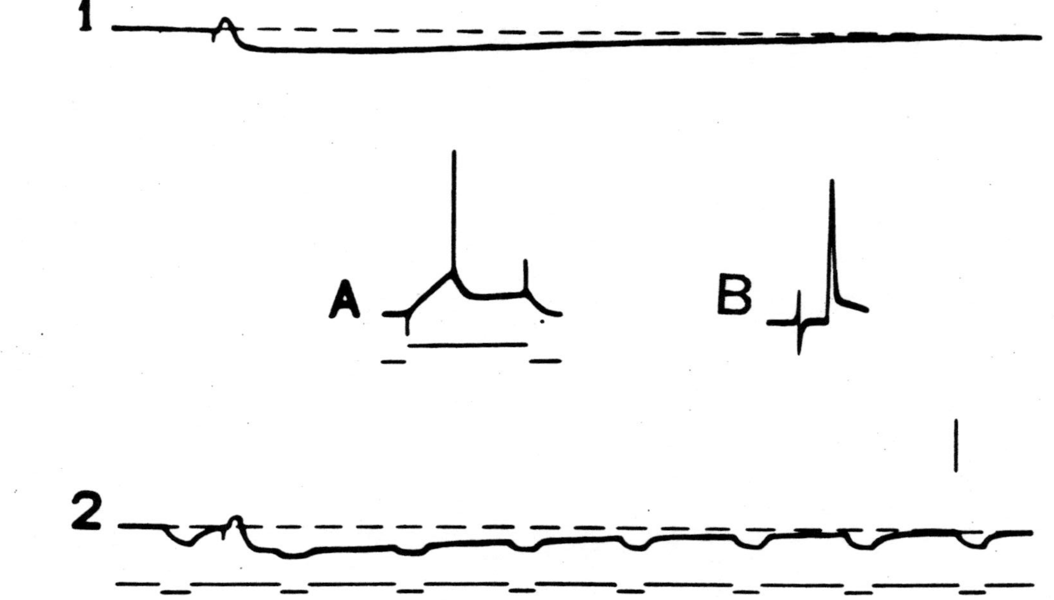

Figure 13. Input resistance changes of a PT neuron during the IPSP evoked by thalamic VPL stimulation: 1. Neuronal response *per se* in a form of EPS-IPSP complex. 2. Membrane resistance tested before and during the neuronal response by means of injecting hyperpolarizing current pulses (1 nA; 20 msec) through the recording microelectrode. Current injection is signalled by the lower trace. Insert A shows a neuronal reaction produced by injection of 1.5 nA depolarizing current pulse of 90 msec duration. Insert B shows identification of the neuron by means of the antidromically elicited spike. Vertical bar represents 20 mV (Z. G. Kokaia, Kokaia, Labakhua and Okujava, 1989).

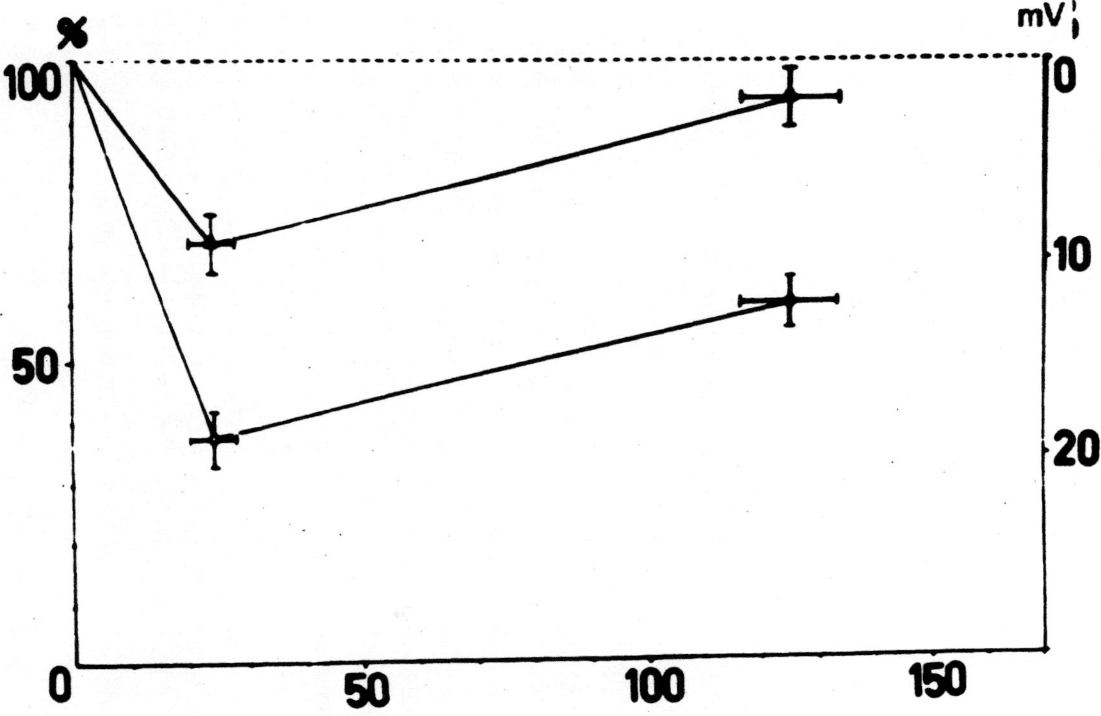

Figure 14. Mean values of input resistance changes (triangles) of 10 PT neurons and the IPSP amplitudes (circles) during early and late components. Left vertical axis represents the input resistance magnitude as a % of the initial value. Right vertical axis, the IPSP amplitude in mV. Horizontal axis indicates time in msec (M. G. Kokaia, Labakhua and Okujava, 1984).

differs from the conductance IPSP by being directly dependent upon metabolism and not involving an increased conductance to ions (Nishi and Koketsu, 1967, 1968; Pinsker and Kandel, 1969; Spencer and Kandel, 1969; Okujava, 1980; Thompson and Prince, 1985). But as will be discussed later in Chapter B.3.d, this latter assumption seems less probable.

c) The effects of artificial polarization upon the IPSPs. Further confirmation of the proposed explanation of the nature of the IPSP components can be obtained in experiments with artificial displacement of the membrane potential by means of depolarizing and hyperpolarizing current injections via the intracellular microelectrode. In Fig. 15 the intracellular record of the sensorimotor cortical neuron is presented. At the initial resting membrane potential level of -60 mV, thalamic VPL stimulation evoked the EPSP with a spike potential followed by the IPSP of 300 msec duration. Artificial hyperpolarization by an applied intracellular current to -64 and -68 mV led to a reduction of the IPSP amplitude. At the level of -72 mV the early component of the IPSP became reversed while the late component, although reduced, still preserved its hyperpolarizing direction. Further hyperpolarization to -76 mV produced an enhancement of the reversed early component but the late component was not reversed. Artificial depolarization to -55 mV produced an increase of the IPSP amplitude. Only in some cases, as in this particular neuron, further extreme depolarization (Fig. 15, - 40mV) led to the decrease of the early component and to the full disappearance of the late one.

Different sensitivity of the early and the late components of the IPSP to the artificial polarization may be the result of different sites of their generation (soma and dendrites). Polarization via the intrasomatically located tip of the microelectrode will exert less

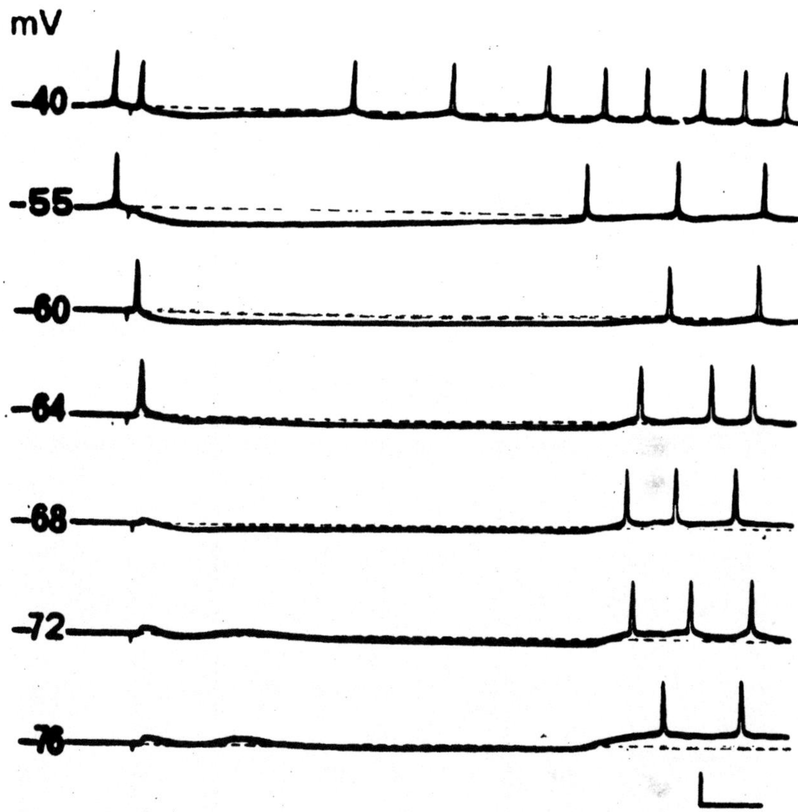

Figure 15. Effect of artificial shift of the membrane potential upon the IPSP elicited by the thalamic VPL stimulation. Initial resting membrane potential level was -60 mV becoming shifted to various levels by means of injecting current of various strength and directions through the recording electrode. Calibration: 30 mV; 40 msec (Z. G. Kokaia, Kokaia, Labakhua and Okujava, 1989).

influence upon the dendrites (Stefanis, 1965; Voronin, 1970; Storozhuk, 1974; Satou, Mori, Tazawa and Takagi, 1982; Avoli, 1986; Serkov, 1986).

As to the depressive action of redundant artificial depolarization upon the IPSPs in certain cases, this fact deserves special explanation, which will be proposed later when considering the probable ionic mechanisms of the IPSPs (Chapter B.4).

3. IONIC MECHANISMS OF THE IPSP GENERATION IN CORTICAL NEURONS

The common mode of synaptic inhibition is accomplished by generation of the conductance IPSPs (Eccles, 1957, 1964). As a matter of fact, the effects produced in the magnitude and direction of the IPSP by varying the initial membrane potential correspond precisely to the changes that would be expected if the current generating the IPSP were due to ions moving down their electrochemical gradients (Eccles, 1964). On the basis of a great number of investigations, the conclusion has been drawn that the interaction between a chemical transmitter substance and a postsynaptic receptor causes an increased conductance to Cl^- or K^+ ions or to their combination in varying degree (Eccles 1957, 1964; Kandel, Spencer and Brinley, 1961; Andersen, Eccles and Loyning, 1964). This type of the IPSP seems to be a common feature of the organization of the mammalian cortex as well (Phillips, 1959; Spencer and Kandel, 1961, 1969; Lux and Klee, 1962; Stefanis and Jasper, 1964a,b; Stefanis 1965; Serkov, 1977, 1986; Z. G. Kokaia, Kokaia, Labakhua, Okujava, 1986). The experimental procedure to test this postulate involves altering the concentration gradient across the postsynaptic membrane for one or other species of ions normally present and, in addition, employing a wide variety of other compounds affecting the definite ionic permeability of the subsynaptic membrane.

For this purpose the procedure of electrophoretic injection of ions via the impaling microelectrode has been widely employed. This method has given valuable information on the ionic mechanisms involved in postsynaptic inhibition.

a) The influence of intracellular injection of chloride upon inhibitory responses. Intracellular injection of chloride into neurons has been tested at various levels of the central nervous system (Coombs, Eccles and Fatt, 1955b; Eccles, 1957, 1964; Curtis, Eccles and Lundberg, 1958; Araki, Ito and Oscarsson, 1961; Kandel, Spencer and Brinley, 1961; Ito, Kostyuk and Oshima, 1962; Andersen, Eccles and Loyning, 1963; Andersen, Brooks and Eccles, 1963; Stefanis, 1965; Voronin, 1970; Serkov, 1977, 1986; Z. G. Kokaia, Kokaia and Okujava, 1986). It appeared that in all such cases IPSPs become depressed and they are readily converted to depolarizing responses even under the influence of mere diffusional and, more so, under the influence of electrophoretic injection of chloride out of an intracellular KC1 microelectrode.

In Fig. 16 an intracellular record of a PT neuron in the cat sensorimotor cortex is shown. Initial response to the thalamic VPL stimulation consists of the EPSP with a spike potential followed by the IPSP. Under the influence of chloride diffusion marked reduction of the IPSP is observed but only its early component becomes converted to depolarization. The predominant effect of intracellular chloride injection on the early component of the IPSP of PT neurons is clearly evident on the averaged graph presented in Fig. 17. Reversal and reduction of the IPSP are statistically significant (p <.01) within the interval of 120 msec beginning from the stimulation artifact.

Quite similar events are observed in intracellular recordings of two non-PT neurons presented in Fig. 18. As is clearly seen on the averaged graph of the IPSPs in 20 non-PT neurons, the chloride injection preferentially reduces and inverts the early component of the inhibitory response; the statistically significant difference between the control and

                                    V.M. Okujava

the experimental points being within the interval of 140 msec from the stimulation artifact (Fig. 19).

As seen in Fig. 20, as a result of long lasting chloride diffusion, the early component may become so strongly converted to depolarization that it generates a high frequency burst of spike potentials with decreasing amplitudes while the late component preserves its hyperpolarizing direction.

b) The influence of intracellular injection of cesium ions upon the inhibitory responses. In addition to chloride, potassium ions may participate in generation of the IPSPs. Regrettably however, this cannot be tested simply by intracellular potassium injection: Since the K^+ ions are normally at such a high level in nerve cells, the ion injection procedures can give but a small increase or decrease in relative concentration, and hence have been indecisive with respect to evidence for or against K^+ ion permeability as a contributory factor in production of the IPSP (Eccles, 1964). Meanwhile an assessment of the relatively considerable contribution of K^+ permeability can be made on the basis of a different experimental approach: on the basis of the effect of potassium channel blockers. According to a number of investigations (Adelman and Senit, 1958, 1966; Benzanilla and Armstrong, 1972; Dubois and Bergman, 1972; Gorman, Woolum and Cornwall, 1982) cesium ions applied to the inner surface of the neuronal membrane exert blocking action upon both voltage-dependent and $Ca2^+$dependent K^+ currents directed outwardly.

As a matter of fact in experiments performed with micropipettes filled with 1 M solution of Cs_2SO_4 for intracellular recording, mere diffusional entry of Cs^+ ions into the neurons is quite sufficient for such changes of electrical events, which presumably must be conditioned by the blockade of potassium channels. As seen in Fig. 21, shortly after penetration of the neuron with such an electrode typical changes of the neuronal activity develop rather rapidly: grouping of action potentials,

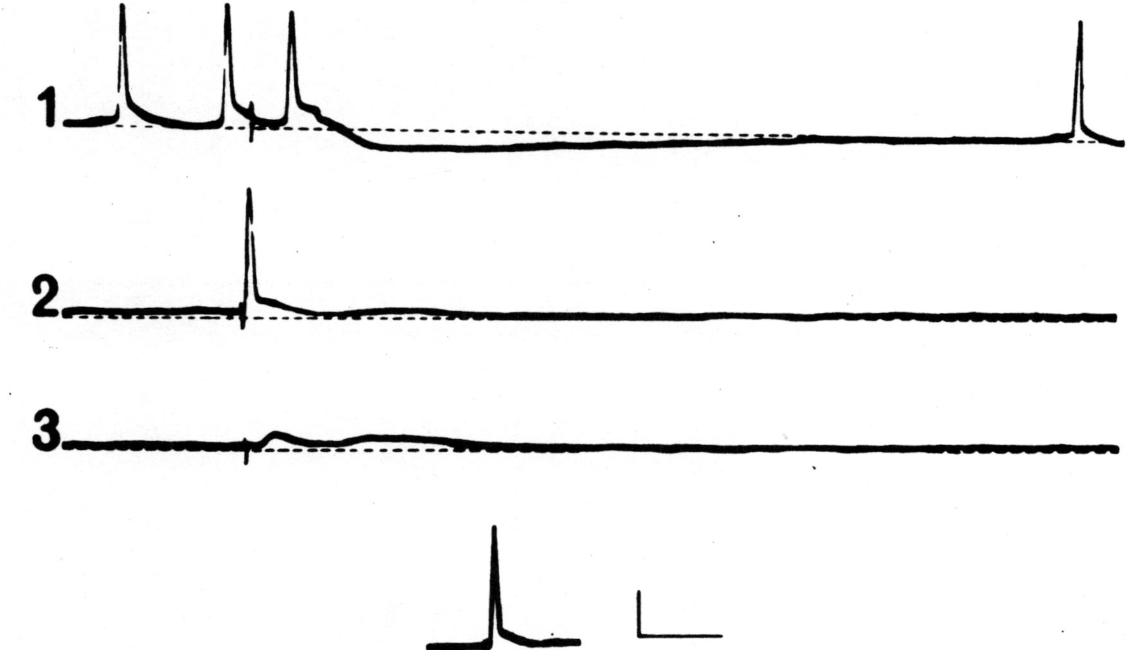

Figure 16. Intracellular records of a pyramidal neuron in the sensorimotor cortex with a potassium chloride-filled microelectrode. 1. Response to a single shock stimulation of the thalamic VPL just after the microelectrode penetration. Responses 2 and 3 were recorded 39 and 91 seconds later, respectively. Calibration: 20 mV; 25 msec (Z. G. Kokaia, Kokaia, Labakhua and Okujava, 1986).

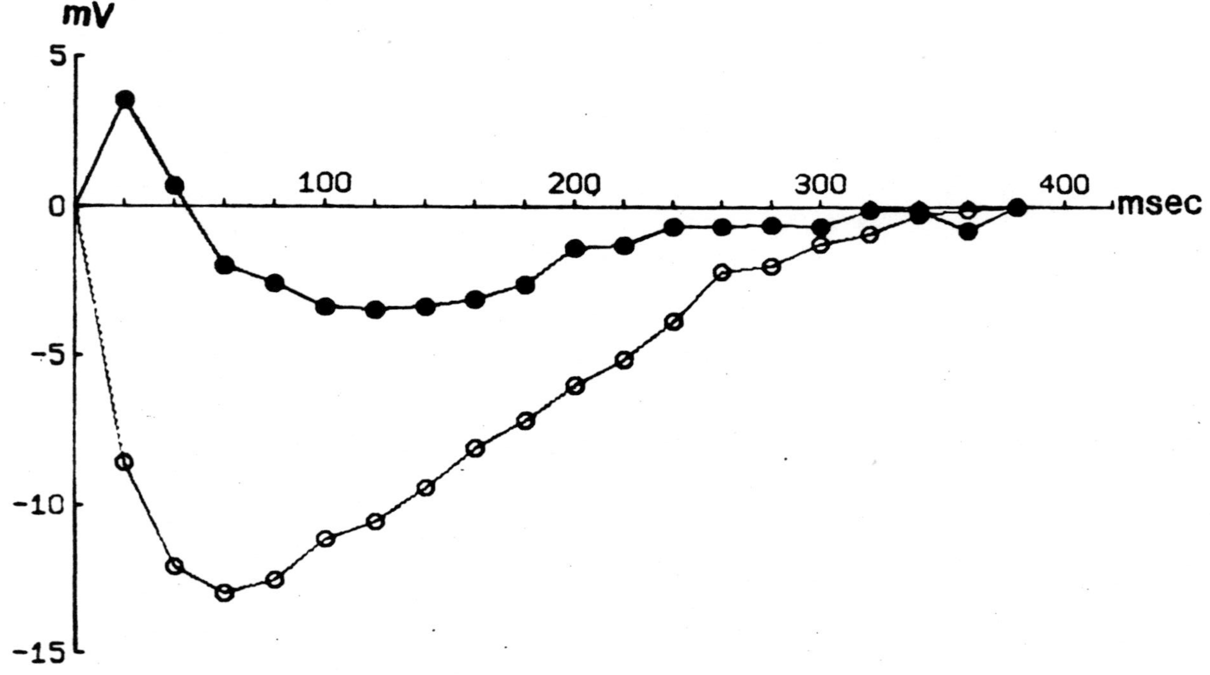

Figure 17. Averaged data demonstrating the influence of intracellular injection of chloride ions upon the IPSPs of pyramidal neurons (n=7). Mean values of the IPSPs before (open circles) and after (closed circles) chloride injection are presented. The difference was statistically significant (p<.01) within the limits of initial 120 msec. Horizontal axis represents time in msec following the VPL stimulation. Vertical axis, mean amplitudes of the IPSPs (Z. G. Kokaia, Kokaia, Labakhua and Okujava, 1986).

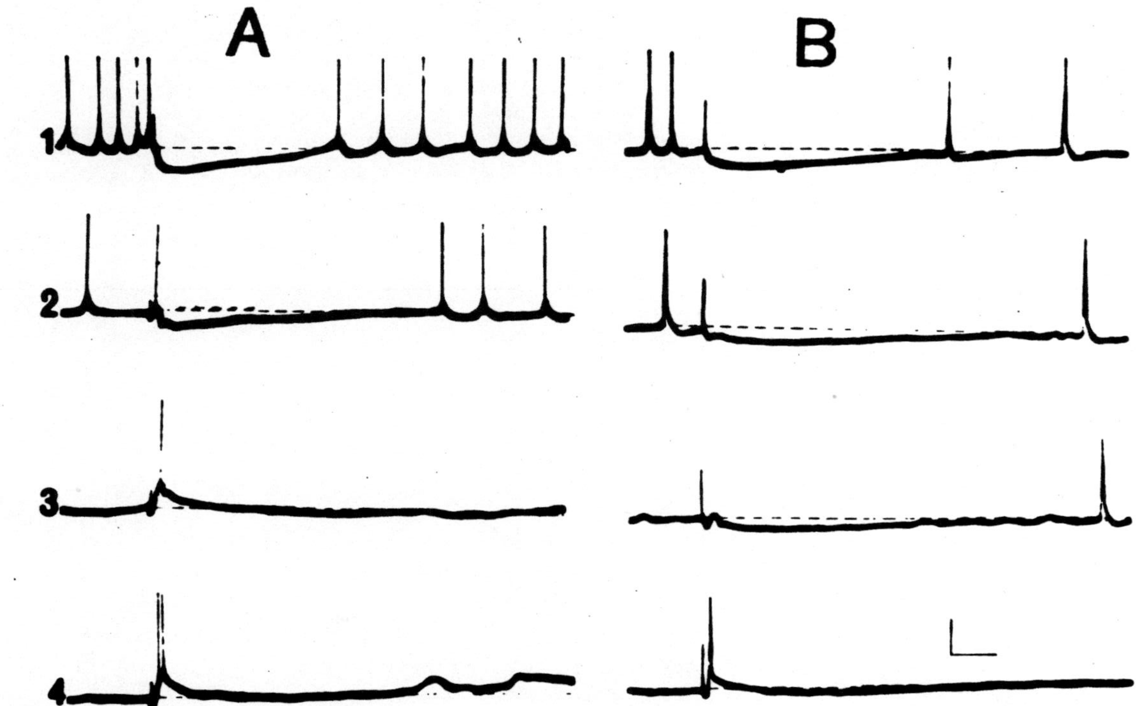

Figure 18. Intracellular records of two nonpyramidal neurons (A and B) with KCl filled microelectrodes. All records show neuronal responses to the thalamic VPL stimulation. First records were carried out immediately following the microelectrode penetration. Records 2, 3 and 4 were performed 13, 175 and 260 sec later in A, and 6, 30 and 60 sec later in B. Calibration: 10 mV; 40 msec (Z.G. Kokaia, Kokaia, Labakhua and Okujava, 1986).

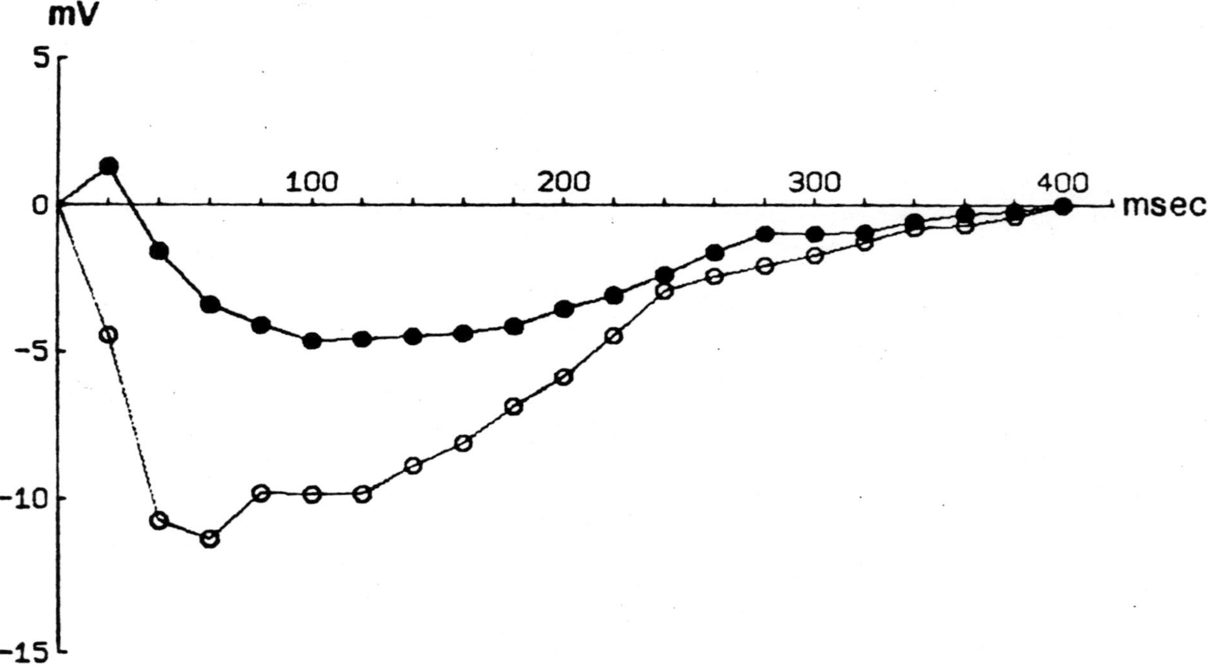

Figure 19. Averaged data demonstrating the influence of intracellular injection of chloride ions upon the IPSPs of non-PT neurons (n=20). Mean values of the IPSPs before (open circles) and after (closed circles) chloride injection. Statistically significant difference was within the limits of initial 140 msec. Horizontal axis represents time in msec following the VPL stimulation. Vertical axis, mean amplitudes of the IPSPs (Z. G. Kokaia, Kokaia, Labakhua and Okujava, 1986).

appearance of bursts divided at first by postburst hyperpolarizations (PBHs), the latter gradually becoming reduced, prolongation of action potentials due to the slowing of their descending phases, and formation of depolarization plateaus with inactivated spikes at their background. In Fig. 22 it is clearly evident that action potentials become prolonged mainly at the expense of the repolarization phase.

At the same time it must be noted that the membrane potential of the neuron does not undergo any significant changes during these procedures. This seems understandable taking into consideration the suggestion (Adelman and Senit 1966; Sjodin, 1966; Franz, Galvan and Constanti, 1986) that Cs^+ ions do not block the channels responsible for creation of the resting potential across the membrane.

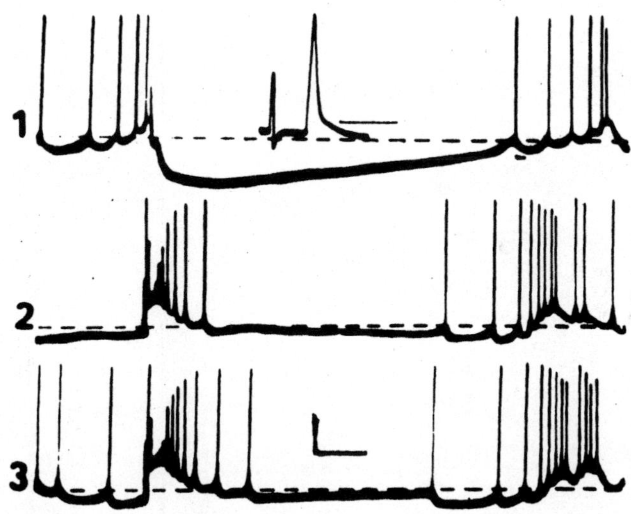

Figure 20. Difference in effects of prolonged diffusion of chloride ions into a PT neuron upon the early and late components of the IPSP. Curve 1 presents a response to the thalamic VPL stimulation in the beginning of recording. Insert shows antidromic identification with the time scale of 10 msec. Records 2 and 3 were carried out 32 and 33 min later. Calibration: 20 mV; 50 msec (Z. G. Kokaia, Kokaia, Labakhua and Okujava, 1986).

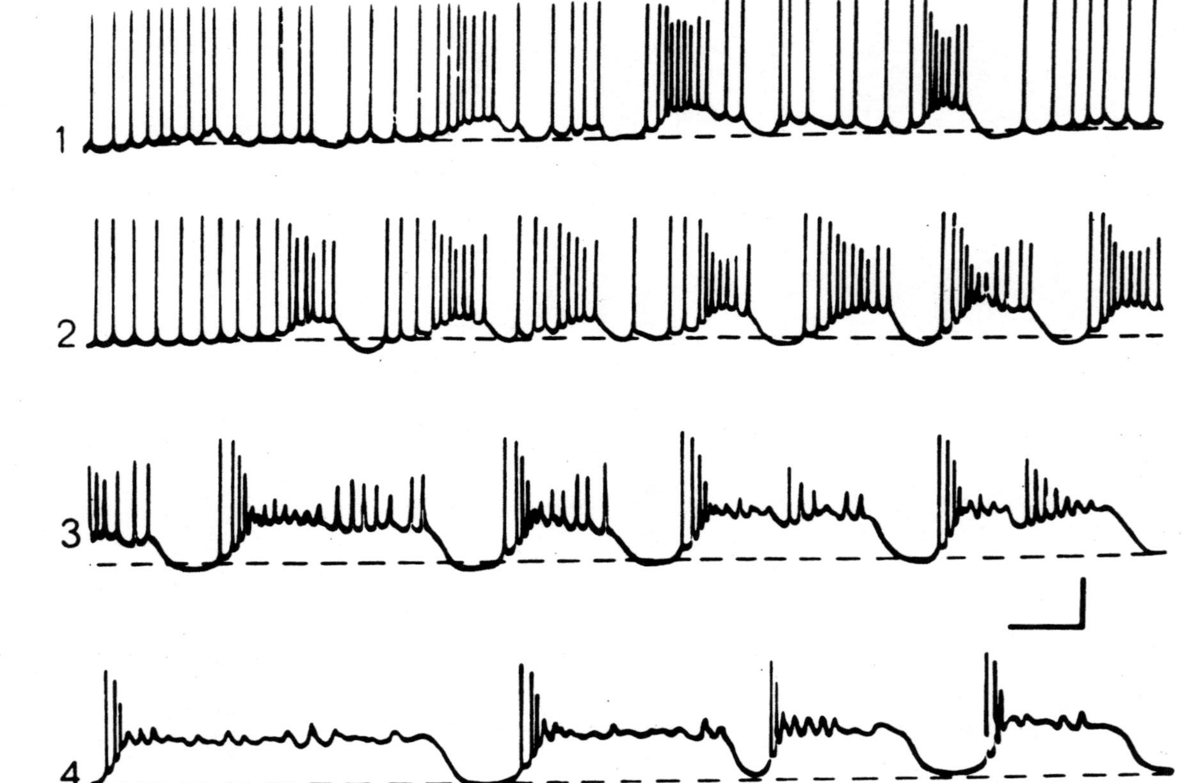

Figure 21. Modifications of spontaneous activity recorded from PT neuron by means of the microelectrode filled with 1 M Cs_2SO_4 solution. Each subsequent record (2, 3, 4) is a direct continuation of the previous one. Calibration: 30 mV; 45 msec (Z. G. Kokaia, Kokaia, Labakhua and Okujava, 1988).

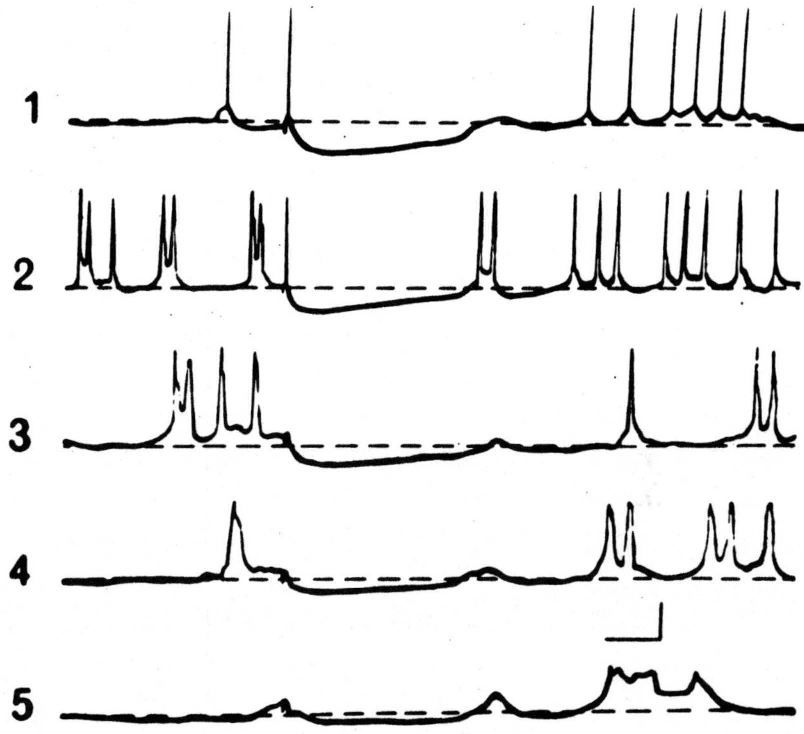

Figure 22. Spontaneous and evoked activity changes in a PT neuron during prolonged intracellular recording with a 1 M Cs_2SO_4-filled microelectrode. Responses were evoked by single electrical stimuli applied to the thalamic VPL. First record shows the response shortly after the microelectrode penetration into the neuron. Records 2 to 5 show responses 1, 3, 4 and 11 min later respectively. Calibration: 20 mV; 50 msec (Z. G. Kokaia, Kokaia, Labakhua and Okujava, 1987).

Transformation of electrical events similar to the above phenomena, namely prolongation of action potentials and appearance of depolarization plateaus, have been found in neurons of neocortical slices in guinea pigs (Franz, Galvan and Constanti, 1986) and in dorsal root ganglia in rats, as well (Mayer, Crunelli and Kemp, 1984), under the influence of intracellular Cs^+ injection. This action of Cs^+ ions is quite comparable to the action of other blockers of voltage dependent K^+ permeability, in particular of tetraethylammonium and Ba^{2+} ions (Krnjevic, Pumain and Renaud, 1971; Franz, Galvan and Constanti, 1986), and Li^+ ions (Mayer, Crunelli and Kemp, 1984). Naturally the blockade of voltage dependent K^+ permeability must hamper the neuronal repolarization leading thus to prolongation of action potentials, reduction of their after-hyperpolarizations and their grouping in a form of bursts, and, finally, to formation of depolarization plateaus. Beside this mechanism the blockade of Ca^{2+} dependent K^+ permeability by Cs^+ ions may also condition the development of depolarization plateaus, because in neurons Ca^{2+} participate in generation of somadendritic spikes in addition to Na^+ (Llinas and Hess, 1976; McAfee and Yorowsky, 1979; Schwartzkroin and Slowsky, 1977; Yoshida, Matsuda and Samejima, 1978; Connors, Gutnick and Prince, 1982; Stafstrom, Schwindt, Chubb, and Crill, 1985; Franz, Galvan and Constanti, 1986), contributing thus to repolarization phenomena, particularly to PBH, which will be discussed at length later in Chapter C.

Intracellular injection of Cs^+ considerably reduces the IPSP amplitude (Fig. 22). In the case presented in Fig. 23 shortening of the IPSP duration is observed after 1.5 min of recording due to the full depression of its late component. As for the early component it remains preserved albeit with reduced amplitude. Still in another case (Fig. 24) considerably greater reduction (by 40%) of the late component in comparison with

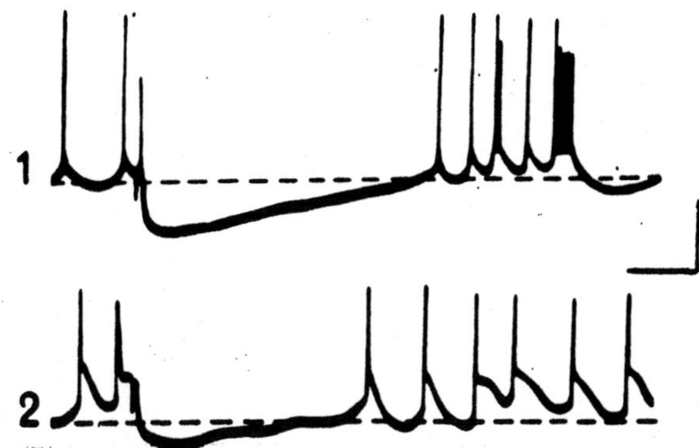

Figure 23. Responses of pyramidal neurons to the thalamic VPL stimulation recorded with a 1M Cs_2SO_4-filled microelectrode shortly after puncturing the cellular membrane (1) and 1.5 min later (2). Calibration. 25 mV; 50 msec (Z. G. Kokaia, Kokaia, Labakhua and Okujava, 1988).

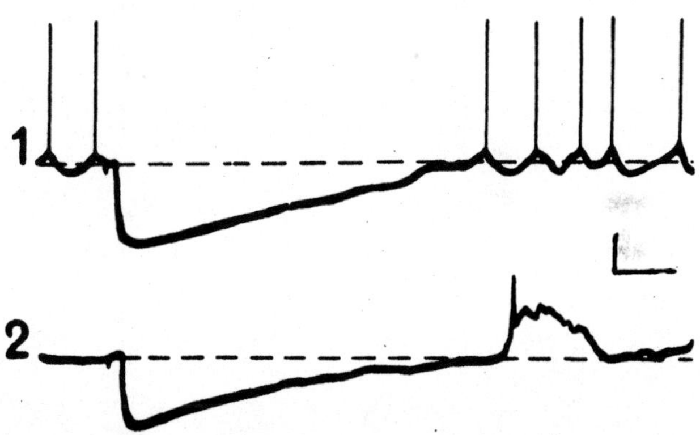

Figure 24. Effect of prolonged (3 min) recording of the PT neuronal hyperpolarization response (1M Cs_2SO_4-filled microelectrode). Responses were elicited by means of single electrical shocks applied to the thalamic VPL. The effect of Cs diffusion is analogous to that shown in Fig. 23. Simultaneous modification of spike activity with formation of depolarization plateaus is observed. Curve 1 shows the beginning of the recording. Curve 2 was recorded 3 min later. Calibration: 10 mV; 50 msec (Z. G. Kokaia, Kokaia, Labakhua and Okujava, 1987).

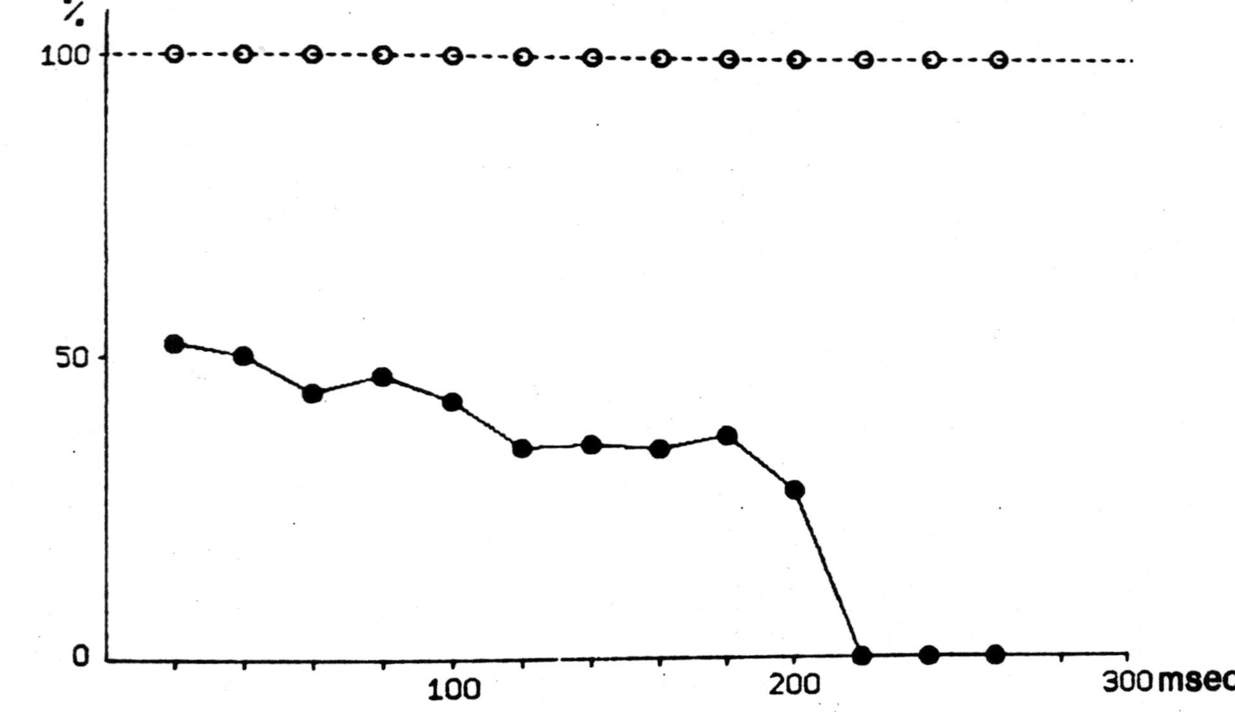

Figure 25. Line graph representing the influence of intracellular infusion of Cs ions upon mean values of the IPSPs in sensorimotor cortical neurons (n=10). Closed circles show the amplitude of the IPSPs after infusion measured at 20 msec intervals as a % of the values of corresponding points in initial recording (open circles) (Z. G. Kokaia, Kokaia, Labakhua and Okujava, 1988).

the early one (by 18%) is seen during recording with a microelectrode filled with 1M Cs_2SO_4 solution. The graph of average values presented in Fig. 25 reveals that notwithstanding the overall reduction of the IPSP under the influence of Cs^+ ions, its early (0-100 msec) and late (120-160 msec) parts differ considerably in their sensitivity, the first one decreasing by 48%-58% and the second one by 66%-100%.

At some lower values of Cs_2SO_4 concentration (0.5M) the IPSP may become similarly reduced while voltage-dependent K^+ channels seem not to be considerably blocked, at least as indicated by the absence of prolonged spikes and depolarization plateaus, probably thus demonstrating the importance of the blockade of $Ca2^+$-dependent K^+ channels (Fig. 26).

In conclusion, on the basis of all those effects of Cs^+ ion injections, particular attention should be given to the possible role of K^+ ions in contributing to the generation of the IPSP as a hyperpolarizing response in cortical neurons.

c) The influence of intracellular injection of ethylenglycol tetroacetic acid upon the inhibitory responses. As was already mentioned above Ca^{2+}dependent K^+ currents may play an important role in generation of hyperpolarization potentials conditioned by the K^+ permeability. Intracellular injection of ethylenglycol tetraacetic acid (EGTA) serves as a proper method of determining the participation of the K^+ permeability changes in the origin of the electrical response under examination, because in cells infected with the calcium chelator, EGTA, a Ca^{2+}-activated K^+ conductance is prevented (Krnjevic, Puil and Werman, 1978; Meech, 1978; Schwartzkroin and Stafstrom, 1980).

In experiments conducted with the use of intracellular microelectrodes filled with 0.5M solution of EGTA it has been established that even mere diffusional entry of the chemical substance into the cortical neuron appears quite sufficient for substantial modification of the IPSP (Z. G. Kokaia, 1987; Z. G. Kokaia, Kokaia,

Labakhua and Okujava, 1987b, 1988). As seen in Fig. 27A, under the influence of EGTA both the early and the late parts of the IPSP, evoked by the thalamic VPL stimulation, become markedly reduced in a non-PT neuron. This effect is more evident in Fig. 27B where the average of five traces is presented. Basically similar effects were obtained in PT neurons (Fig.28).

The averaged data concerning the EGTA influence upon the IPSP in 12 sensorimotor cortical neurons are graphically presented in Fig. 29. The late part (240-420 msec) of the IPSP is diminished more markedly (by 44%-62%) than the early part (0-220 msec by 36%-40%). Thus, the results of Cs^+ and EGTA injections confirm the impotent role of a K^+ potential in the origin of hyperpolarizing inhibitory responses in cortical neurons, this K^+ potential being predominant in the late component of the response and to a substantial extent being conditioned by a $Ca2^+$-activated K^+ conductance. Synaptically activated K^+ potentials may probably also contribute to the development of the IPSP (note dissimilar efficiencies of Cs^+ and EGTA influences, Fig. 25 and Fig. 29).

d) The influence of intracellular injection of vanadate upon the inhibitory responses. According to a considerable amount of literary data (Nishi and Koketsu, 1967, 1968; Pinsker and Kandel, 1969; Spencer and Kandel, 1969; Okujava, 1980; Thompson and Prince, 1986), in addition to the common mode of chemical synaptic inhibition, the conductance IPSPs, a less common mode of chemical inhibition, electrogenic pump IPSPs may make an important contribution to formation of long-duration hyperpolarization potentials. This mode of chemical inhibition differs from the conductance IPSP by being directly dependent upon metabolism and not involving an increased conductance to ions. In such a case the chemical

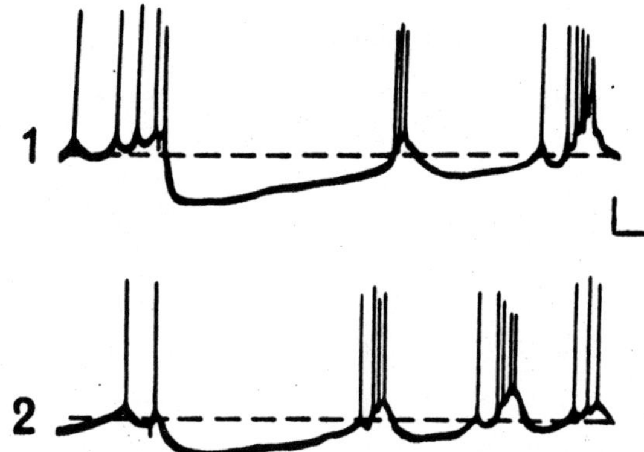

Figure 26. Responses of a sensorimotor cortical neuron to single shock stimulation of the thalamic VPL nucleus are shown in the beginning of recording (1) with a 0.5 M Cs_2SO_4-filled intracellular microelectrode, and 11 min later (2). Calibration: 10 mV; 50 msec (According to studies performed with Z. G. Kokaia, 1987).

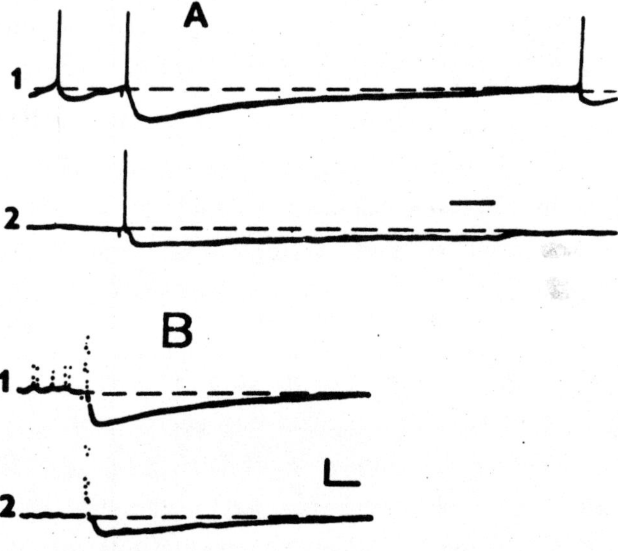

Figure 27. Effect of intracellular infusion of EGTA upon the IPSP in the response of a non-PT neuron to the thalamic VPL stimulation. In A single curves are shown in the beginning of recording (1), and 6 min later (2). In B same responses are presented in a form of the average of five sweeps. Calibration: 10 mV; 50 msec (According to studies performed with Z. G. Kokaia, 1987).

transmitter can activate an electrogenic Na^+ pump which produces a flow of Na^+ against its concentration gradient, thereby generating a hyperpolarizing potential. The active process involved in generating this IPSP appears to be a Na^+-K^+-activated ATPase identical to that, which maintains the normal Na^+ and K^+ distribution across nerve cell membranes (Skou, 1965) and, thus, maintains the membrane potential. Therefore the absence of membrane resistance changes during such an IPSP is its typical feature.As far as the membrane resistance during the late component of the IPSP returns to the background value, the contribution of the electrogenic pump IPSP to the origin of cortical inhibitory responses must be considered. For the sake of testing such a possibility the intracellular injection of vanadate has been used in the author's laboratory (Z.G. Kokaia, 1987). Vanadate acts as a potent Na^+-K^+-ATPase inhibitor when applied to the inner surface of the membrane (Cantley, Cantley and Josephson, 1978; Cantley and Aisen, 1979; Jandhyala and Hum, 1983).

As demonstrated in Fig. 30 and Fig. 31, neither in PT neurons, nor in nonPT ones is there any noticeable modification of the IPSPs following the intracellular vanadate infusion. Particularly demonstrative is the graph of the averaged IPSP amplitudes before and after the vanadate injection (Fig. 32). The possibility of insufficient flow of vanadate into the cell for blocking the Na^+-K^+activated ATPase seems rather improbable in these experiments, because very low vanadate concentration (5×10^{-6} M) was needed in cerebral cortical homogenates of rats and mice for full inhibition of such activity (Krivanek, 1981; Folbergrova, 1986). Sizeable concentration gradient directed into the cell (recording micropipette was filled with 10 mM solution of Na_3VO_4), long duration of recording (up to 20 min), as well as application of negative currents to the microelectrode for vanadate injection during several minutes (vanadate was present in the solution in a form of negatively charged VO_4^{3-}) allowed assuming that a sufficient quantity of

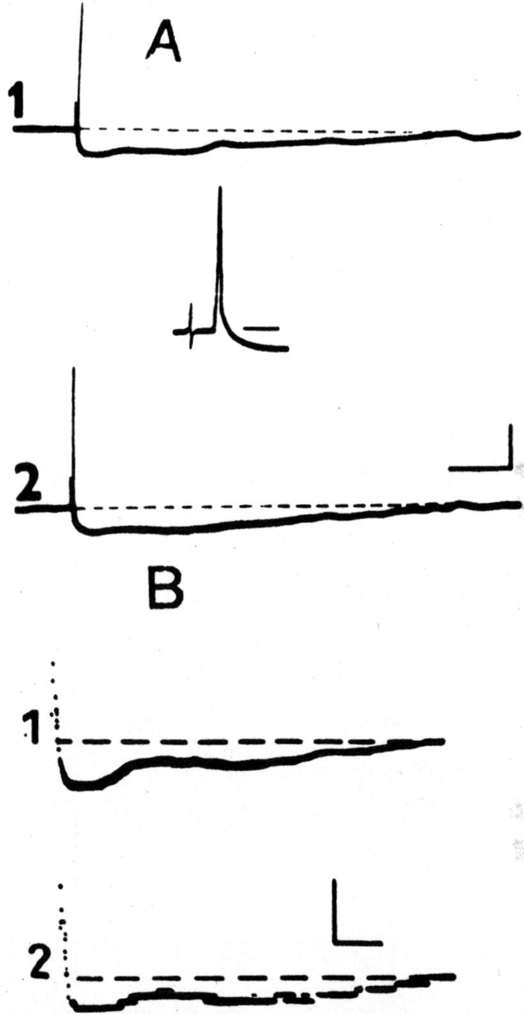

Figure 28. Effect of intracellular EGTA infusion upon the PT neuronal IPSPs elicited by the VPL stimulation. In A single curves are presented in the beginning of recording (1) and 13 min later (2). Insert shows identification test with the time scale of 5 msec. Vertical bar represents 20 mV; horizontal bar, 50 msec. In B same responses in a form of the average of eight sweeps are shown. Vertical bar represents 10 mV, horizontal bar, 30 msec (Z. G. Kokaia, Kokaia, Labakhua and Okujava, 1988).

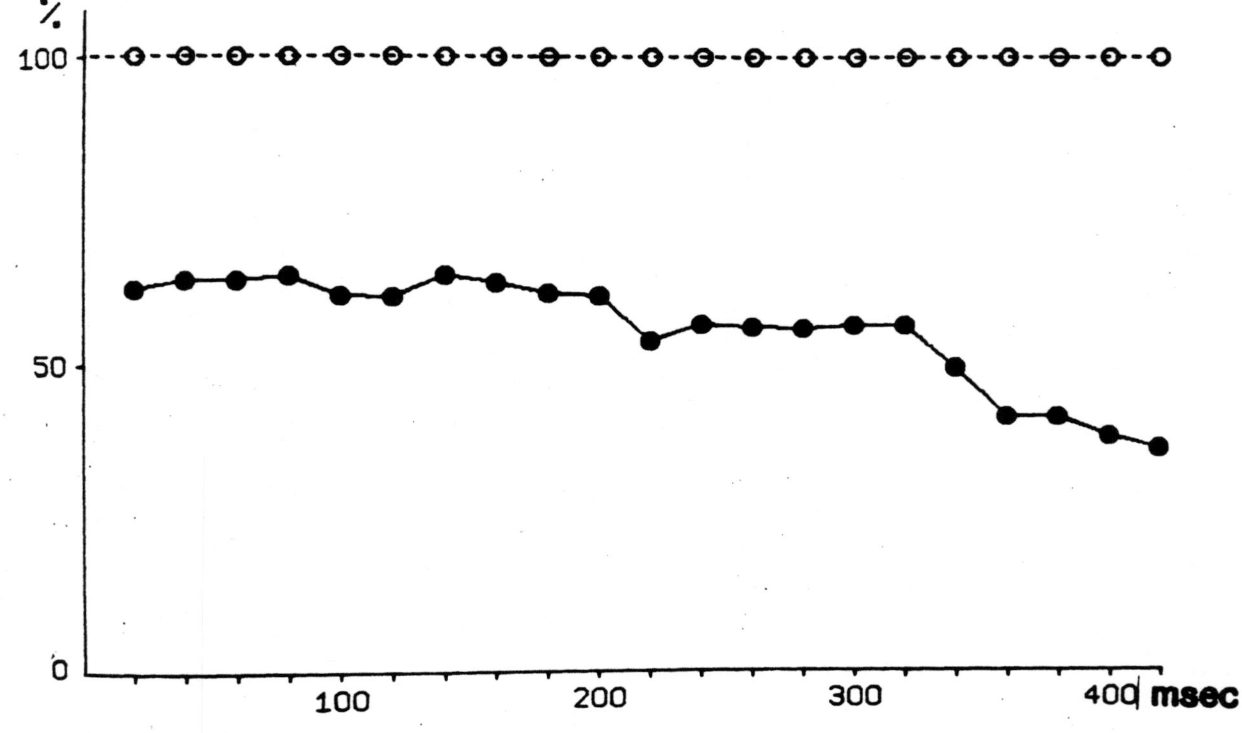

Figure 29. Line graph representing the influence of intracellular infusion of EGTA upon mean values of the IPSPs in PT neurons (n=12). Closed circles show the amplitude of the IPSPs measured after infusion at 20 msec intervals as a % of the values of corresponding points in initial recording (open circles) (Z. G. Kokaia, Kokaia, Labakhua and Okujava, 1988).

Figure 30. PT neuronal responses to the thalamic VPL stimulation before (1) and after (2) intracellular vanadate injection. Calibration: 20 mV; 50 msec (According to studies performed with Z. G. Kokaia, 1987).

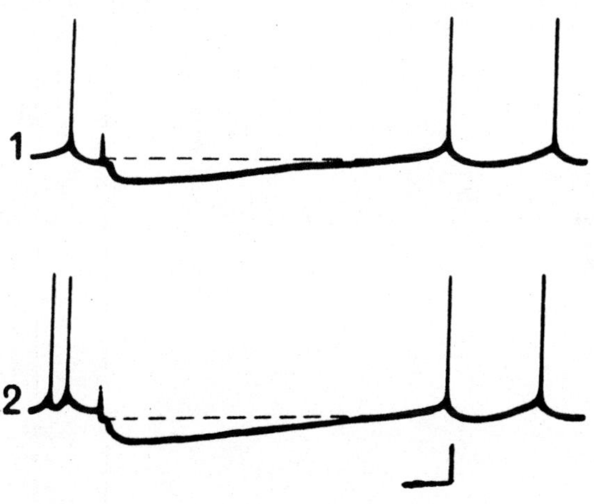

Figure 31. Non-PT neuronal responses to the thalamic VPL stimulation before (1) and after (2) intracellular vanadate injection. Calibration: 10 mV; 100 msec (According to studies performed with Z. G. Kokaia, 1987).

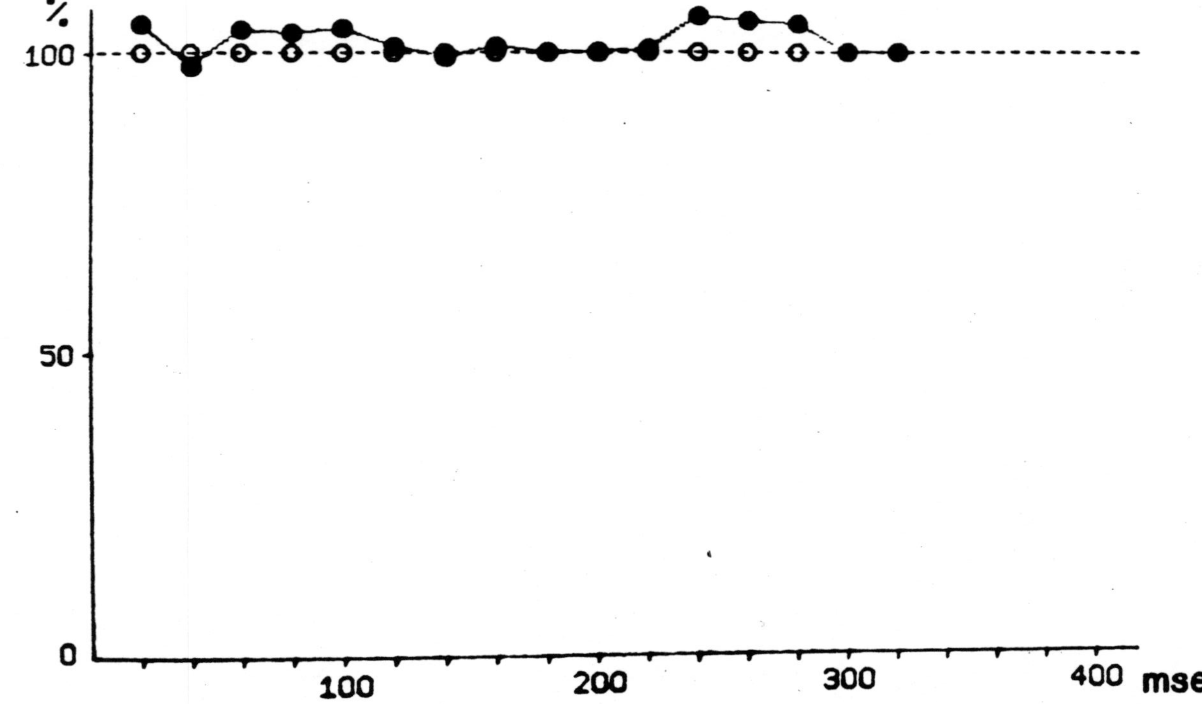

Figure 32. Line graph representing mean values of the IPSPs in sensorimotor cortical neurons (n=12) evoked by the thalamic VPL stimulation before (open circles) and after (closed circles) intracellular vanadate injection (According to studies performed with Z. G. Kokaia, 1987).

vanadate should be entering the cell in this situation for blocking the Na^+-K^+-ATPase to a considerable extent. Therefore it may be concluded that the electrogenic pump does not contribute significantly to formation of any component of the described cortical IPSPs. Although its participation in longer and later hyperpolarizing events, e.g. in postetanic hyperpolarization in snail neurons (Kostyuk and Krishtal, 1981) and in postictal hyperpolarization following the seizure activity in cortical neurons (Okujava, 1980), is not ruled out.

4. CONCLUSION

Summing up all the experimental results stated above, it may be concluded that rather convincing evidence has been accumulated to demonstrate composite nature of the long-lasting IPSPs, which are readily evoked in cortical neurons in response to either peripheral sensory stimulation or to stimulation of various cerebral structures. These IPSPs seem to consist of two components—the early and the late ones with different pharmacological properties, differing sites of origin and diverse ionic mechanisms. It must be noted that no clear-cut boundary exists between the components; the duration of each one depends upon the duration of the whole compound IPSP and they partially overlap each other. It may be assumed that conventionally the early and the late components correspond to the first and the second halves of the IPSP.

Investigation of the sensitivity of the IPSP to the blocking action of strychnine revealed pharmacological a distinction between its early and late components, the former being highly sensitive to strychnine while the latter appearing relatively strychnine-resistant.

The membrane resistance measurements during the IPSP gave evidence in favour of diverse sites of origin of the early and the late components: the early component must be predominantly generated in the somatic region of the neuron while the later component must be evoked in more remote sites (probably in proximal parts of dendrites). The alternative explanation of the origin of the late component as a kind of a nonconductance electrogenic pump IPSP seems rather improbable on the basis of the effects of the intracellular vanadate injection.

Moreover, the supposition about diverse generation sites of the IPSP components has been confirmed by the results of artificial changes of the membrane potential. The initial component appeared extremely sensitive to such changes increasing gradually with depolarization and decreasing with hyperpolarization of the membrane. Particularly noteworthy is the fact that only this component became inverted at a definite level of hyperpolarization. As to the paradoxical fact of reduction of the IPSP discovered in some cases of extreme artificial depolarization, this phenomenon might be the result of activation of certain channels which generate subthreshold currents and modify the input resistance of the neuron, e.g., a persistent subthreshold Na^+ current observed during depolarization in cat neocortical neurons *in vitro* (Stafstrom, Schwindt, Chubb and Crill, 1985). Activation of this event may lead to the reduction of the IPSP amplitude due to the shunting of the IPSP current via the Na^+ channels, on the one hand, and, on the other hand, due to the fact that having an opposite direction this current diminishes the overall magnitude of the current, which conditions the IPSP.

In this way, the findings presented here serve to illustrate that the early component of the cortical IPSP must be predominantly conditioned by the activity of axo-somatic inhibitory synapses while the late component must be mainly generated by axo-dendritic ones (presumably located upon the proximal parts of dendrites).

In order to account for all the above effects of ions and various chemical injections on the IPSP it has been postulated that increased permeabilities to Cl^- and K^+ participate in generation of the IPSP, each one having a different share in formation of its early and late components. An increase in the internal Cl^- concentration by injecting chloride ions into the interior of a neuron, using a current or simply by diffusion, revealed predominant sensitivity of the early components to the concentration gradient of this particular ion. If enough Cl^- was injected, the early component would appear drastically altered. It was extremely diminished or even reversed at resting potential. At the same time it must be noted that the influence of chloride injection was not so evenly obvious throughout the full duration of the IPSP becoming less evident during the late component, never leading to its reversal. Thus, one may conclude that the contribution of the chloride channels is most prominent in the origin of the somatic part of the IPSP, their participation diminishing with a distance. An alternative explanation, that during intracellular injection chloride ions do not penetrate into dendrites in sufficient amounts to produce an essential increase in the internal Cl^- concentration for the IPSP reversal, seems rather improbable, because in proximal shafts of dendrites diffusion appears quite unlimited. For instance, intracellularly injected Lucifer Yellow freely penetrates even into distal parts of dendrites (Lancaster and Wheal, 1984). Moreover, as already documented above, intracellularly injected Cs^+ and EGTA exert their preferential influence upon the late component of the IPSP demonstrating thus that they have reached the site of generation of this component.

In addition, intracellular injection of the potassium channel blocker, Cs^+, convincingly revealed an increased K^+ ion permeability as a contributory factor in production of the IPSP. The contribution from K^+ ion permeability appears relatively less important during the early component, becoming predominant during the later one.

Consequently it may be suggested that the ionic mechanism producing the IPSPs consists of the increased permeability in the postsynaptic membrane to both K^+ and Cl^- ions under the influence of the inhibitory transmitter, each of these permeabilities being expressed to different degrees at axo-somatic and axo-dendritic synapses.

At the same time, speaking about hyperpolarization produced by K^+ currents, a special type of K^+ permeability—Ca^{2+}-activated K^+ conductance, must be particularly considered (Meech and Strumwasser, 1970; Krnjevic and Lisiewicz, 1972; Meech, 1972, 1978; Meech and Standen, 1975; Krnjevic, Puil and Werman, 1978; Gorman, Woolum and Cornwall, 1982). Intracellular injection of the calcium chelator EGTA has contributed a great deal to the recognition of this type of K^+ conductance (Meech, 1978; Krnjevic, Puil and Werman, 1978; Schwartzkroin and Stafstrom, 1980). The results of similar experiments, presented above, served to clearly illustrate that K^+ permeability under examination to a substantial extent was Ca^{2+}-dependent.

The mechanisms of activation of Ca^{2+}-dependent K^+ permeability are not fully understood. According to the most probable explanation, Ca^{2+} ions, entering the cell exert their influence via calmodulin. Calmodulin has been shown to be a calcium receptor protein abundantly present in nervous tissue, with a strong and specific binding affinity for calcium (Klee, Crouch and Richman, 1980). Under the influence of calcium it undergoes a conformational change and turns into an activated state. Ca^{2+}-calmodulin complex can activate adenilate cyclase leading thereby to membrane protein phosphorylation by means of cAMP-dependent proteinkinases, or it can directly modify the membrane proteins by means of Ca^{2+}-calmodulin-dependent protein kinases (Nestler, Waleas and Greengard, 1984).

Under the influence of Ca^{2+} ions released from cytoplasmic sources Ca^{2+} dependent K^+ channels can become activated (Trautman and Marty, 1984), but, as a rule, such conductance is mostly activated due to the Ca^{2+} entry into the cell via voltage-dependent calcium channels. It is well known that, besides Na^+ ions, Ca^{2+} ions, too, take part in the development of somadendritic action potentials (Oomura, Ozaki and Maeno, 1961; Llinas and Hess, 1976; Schwartzkroin and Slowsky, 1977; O'Lague, Potter and Furshpan, 1978; Akasu and Koketsu, 1981; Hagiwara and Byerly, 1981; Kostyuk and Krishtal, 1981; Suppes, 1984; Stafstrom, Schwindt, Chubb and Crill, 1985; Franz, Galvan and Constanti, 1986). In addition, it must be noted that Ca^{2+} ions can enter the cell not solely through specific Ca^{2+} channels but partially through Na^+ channels as well, due to incomplete selectivity of the latter (Kostyuk and Krishtal, 1981).

Moreover, the voltage-dependent Ca^{2+} and Na^+ channels are not the only way for calcium influx. These ions readily move through chemosensitive Ca^{2+} channels activated by neurotransmitters. This phenomenon has been shown in motoneurons of the frog (Buhrle and Sonnhof, 1983), as well as in hippocampal pyramidal cells (Nicoll and Alger, 1981) and in neurons of the rat motor cortex (Pumain and Heinemann, 1985). Excitatory neurotransmitters produce also intracellular influx of Ca^{2+} ions in neuromuscular junctions of both vertebrates and invertebrates (Takeuchi 1963; Onodera and Takeuchi, 1975; Anwyl, 1977; Miledi, Parker and Schalow, 1980).

The supposition about the influx of Ca^{2+} ions under the action of excitatory neurotransmitters without participation of action potentials to some degree, has been confirmed by our own data presented earlier in this volume; intracellularly injected Cs^+ and EGTA produced reduction of the IPSP amplitude even in cases when the IPSP was proceded solely by an EPSP without an accompanying action potential. Therefore, it may be admitted that during the EPSP chemosensitive Ca^{2+} channels became activated and Ca^{2+} ions entering the neuron evoked hyperpolarizing Ca^{2+}-dependent K^+ currents. Such EPSPs sometimes can be even concealed by the earliest part of the IPSP as revealed by the unmasking effect of the iontophoretic application of strychnine (M. G. Kokaia, Labakhua and Okujava, 1984).

The inhibitory hyperpolarizing potential of PT neurons appear to be qualitatively similar to those of non-PT cells. PT neurons are the pyramidal cells with corticofugal axons that project into the pyramidal tract. It is possible to identify these cells by stimulating their axons in the peduncle or the pyramid and antidromically discharging them. Thus only a small number of effector neurons fall into this group while the group of non-PT neurons appears much vaster and more heterogenous (Chizhenkova, 1986). Extrapyramidal effector cells constitute a significant part of non-PT neurons. Such effector neurons have considerably larger size in comparison with interneurons (Chizhenkova, 1986). Consequently they must be much more accessible for intracellular recording. Thus the similarity concerns the electrophysiological and pharmacological properties of the pyramidal and the extrapyramidal effector cells mainly.

C. OTHER TYPES OF HYPERPOLARIZING POTENTIALS IN CORTICAL NEURONS AND THEIR INTERACTION WITH THE IPSPS

It will be noted that in the central nervous system inhibitory synapses producing IPSPs optimally control the generation of impulse discharges in neurons (Eccles, 1964). In addition to such postsynaptic events some other hyperpolarizing potentials of a non-synaptic nature play an important role in controlling the neuronal firing. First of all, the after-potentials following a spike must be considered. As discovered in spinal motoneurons (Coombs, Eccles and Fatt, 1955a) the spike in the soma-dendritic membrane is followed by a large and prolonged after-hyperpolarization (AHP), while the after-potentials are negligible in the axon hillock membrane as shown by blocking the soma-dendritic spike and recording from the some or, occasionally, by recording from the axon hillock region. This large somatic AHP is thought to be a major factor in limiting the firing frequency of motoneurons, because it is associated with a substantial subnormal period (Willis and Grossman, 1973). According to the majority of investigations the AHPs following solitary action potentials must be conditioned by enhanced K^+ conductance (Barrett, Barret and Crill, 1980; Gustaffson and Wigstrom, 1981a; Deschenes, Paradis, Roy and Steriade, 1984; Nohmi and Kuba, 1984; Stafstrom, Schwindt, Flatman and Crill, 1984; Tokimasa, 1984; Fowler, Greene and Weinreich, 1985; Fowler, Wonderlin and Weinreich, 1985; Hill, Arhem and Grillner, 1985; Van Dongen, Grillner and Hokfelt, 1985; Yoshimura, Polosa and Nishi, 1986). In general, the AHP may be explained as a uniform process caused either by voltage-dependent potassium conductance responsible for repolarization phase of an action potential (Hodgkin and Huxley, 1952; Hodgkin and Keynes, 1955; Fournier and Crepel, 1984) or by Ca^{2+}-activated K^+ conductance (Krnjevic, Puil and Werman, 1978; Brown and Griffith, 1983; Deschenes, Paradis, Roy and Steriade, 1984; Leonard and Wickelgren, 1985; Segal and Barker, 1986). However a multitude of experimental evidence has been accumulated indicating that the AHP consists of two dissimilar phases differing from each other in duration and pharmacological sensitivity (Cherubini, North and Surprenant, 1984; Nohmi and Kuba, 1984; Tokimasa, 1984; Fowler, Greene and Weinreich, 1985; Fowler, Wonderlin and Weinreich, 1985; Hill, Arhem and Grillner, 1985; Van Dongen, Grillner and Hokfelt, 1985; Yoshimura, Polosa and Nishi, 1986). According to Stafstrom, Schwindt and Crill (1984), in sensorimotor cortical slices the initial fast phase is conditioned by activation of the rapid voltage-dependent K^+ permeability responsible for the repolarizing part of the action potential, while the late phase must be the result of activation of slow Ca^{2+}-dependent K^+ channels. In fact, as already mentioned in Chapter B.3.b, intracellular injection of Cs^+ ions, which exert blocking action upon potassium currents, leads to prolongation of action potentials at the expense of their repolarization phase, and to drastic reduction of their AHPs (Fig. 20 and Fig. 21), thus producing the grouping of action potentials in a form of bursts. Consequently the paramount contribution of K^+ permeability to the generation of AHPs seems quite probable. Intracellular infusion of EGTA considerably decreases the AHP amplitude (Fig. 33) revealing the role of Ca^{2+}-dependent K^+ currents in the origin of this hyperpolarization. It must be noted that the short initial part of the AHP (first 20 msec) does not undergo any marked alteration while the later part becomes considerably reduced (by 30%-100%), as is clearly seen in Fig. 34 where mean values of AHP of five neurons are graphically presented. Therefore it may be concluded that the earliest part of the AHP is conditioned by voltage-dependent K^+ permeability, which is responsible for the repolarizing phase of the action potential, while the later and more prolonged part of the AHP must be conditioned by K^+ conductance activated by Ca^{2+} entering the cell during the action potential either through Na^+ channels (Kostyuk and Krishtal, 1981) and/or through the voltage-dependent Ca^{2+} channels (Llinas and Hess, 1976; Schwartzkroin and Slowsky, 1977; Hagiwara and Byerly, 1981; Connors, Gutnick and Prince, 1982; Stalstrom, Schwindt, Chubb and Crill, 1985; Franz, Galvan and Constanti, 1986).

In such a way the conclusion seems justified that a major factor in limiting the neuronal firing frequency (Kernell, 1965; Kernell and Sjoholm, 1973; Krnjevic, Puil and Werman, 1978; Barrett and Barrett, 1976; Alvarez-Leefmans and Miledi, 1980; Gustaffson and Wigstrom, 1981a,b; Baldissera and Gustaffson, 1974; Baldissera, Campadelli and Piccinelli, 1983; Yoshimura, Polosa, Nishi, 1986; Stafstrom, Schwindt and Crill, 1984a,b; Stafstrom, Schwindt, Chubb and Crill, 1985) consists of two phases: the short initial phase conditioned by voltage dependent K^+ conductance and the later one conditioned by Ca^{2+}-dependent K^+ permeability. At the same time it will be very important to test the possibility of participation of inhibitory synaptic influences in generation of the hyperpolarization succeeding the action potential, because recurrent inhibitory synaptic effects can be clearly demonstrated in Betz cells of the motor cortex during the antidromic stimulation of the pyramidal tracts, suggesting that the IPSPs, in these cases, are most probably mediated through recurrent axon collaterals (Phillips, 1956, 1959; Stefanis and Jasper, 1964; Humphrey, 1968).

If the IPSP appears to take part in formation of hyperpolarization following the action potential then it must be concluded that when the cortical cell is excited its nervous impulse by means of recurrent collaterals feeds back onto the parent neuron presumably via an internuncial neuron or a series of interneurons, finally exerting inhibitory transsynaptic action upon the very initial cell. This supposition can be decisively verified by means of intracellular injection of Cl⁻ ions as far as the origin of the IPSP is definitely conditioned by C1⁻ conductance. Although the effect of this procedure appeared far less consistent than the influence of Cs^+ and EGTA injections, in certain cases (Li, Okujava and Bak, 1971) the hyperpolarization was converted to a depolarization. Fig. 35 is an example showing the responses to depolarizing current of 2.0×10^{-9} A and varying durations in a range of 12-72 msec. As the current was terminated, the depolarizing response was followed by hyperpolarization. In this experiment, after 25 such responses were obtained, negative current of 0.75×10^{-9}A was passed to the cell through the recording micropipette for 2 min. Then, depolarizing currents of similar strength and duration were applied. The responses, now (Fig. 35 B), consisting of initial depolarization and spike discharges, were not changed; but the subsequent hyperpolarization tended to give place to depolarization.

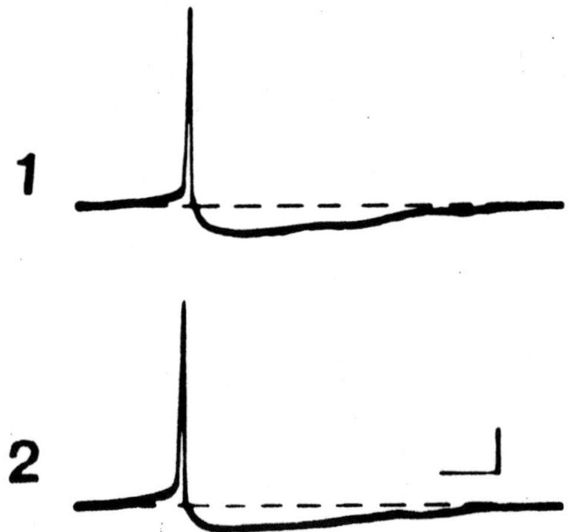

Figure 33. AHP following a spontaneous spike potential in a PT neuron in the beginning of recording with an EGTA-filled intracellular microelectrode (1), and 3 min later (2). Calibration: 10 mV; 10 msec (Z. G. Kokaia, Kokaia, Labakhua and Okujava, 1988).

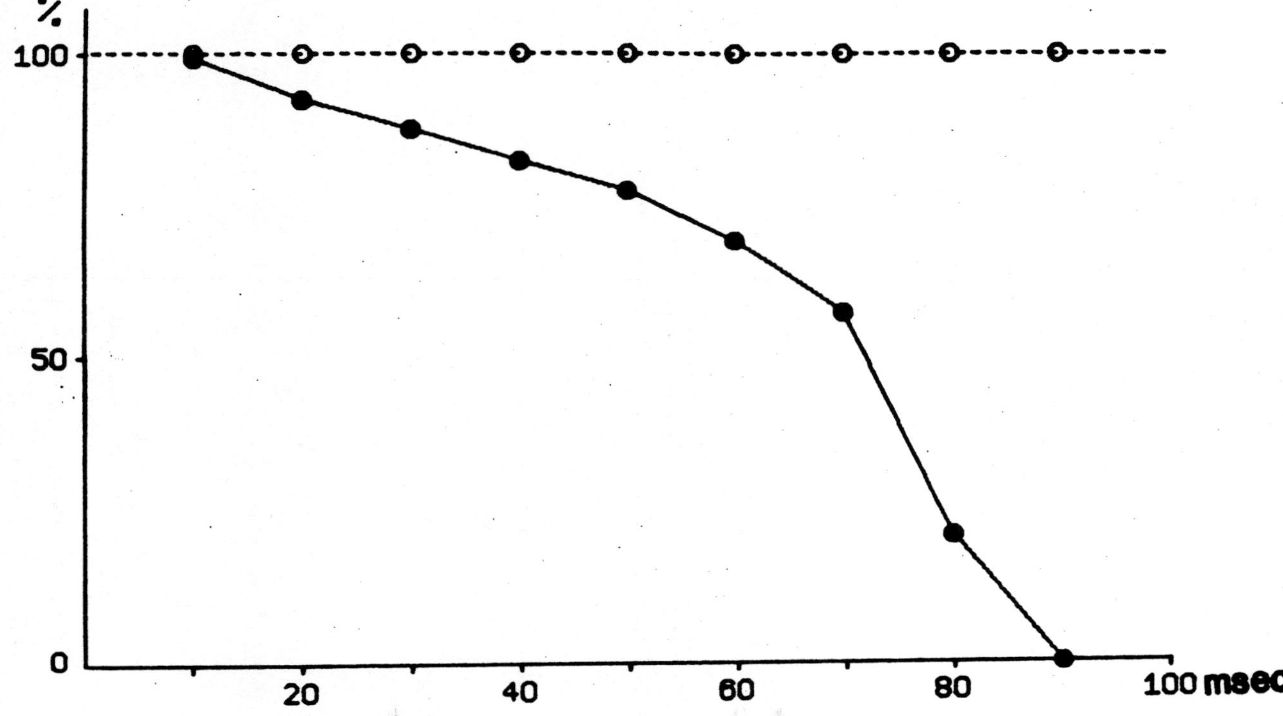

Figure 34. Mean values of AHP in sensorimotor cortical neurons (n=5) after EGTA injection (closed circles) measured at 20 msec intervals as a % of the values at corresponding points in initial recording (open circles) (Z. G. Kokaia, Kokaia, Labakhua and Okujava, 1988).

V.M. Okujava

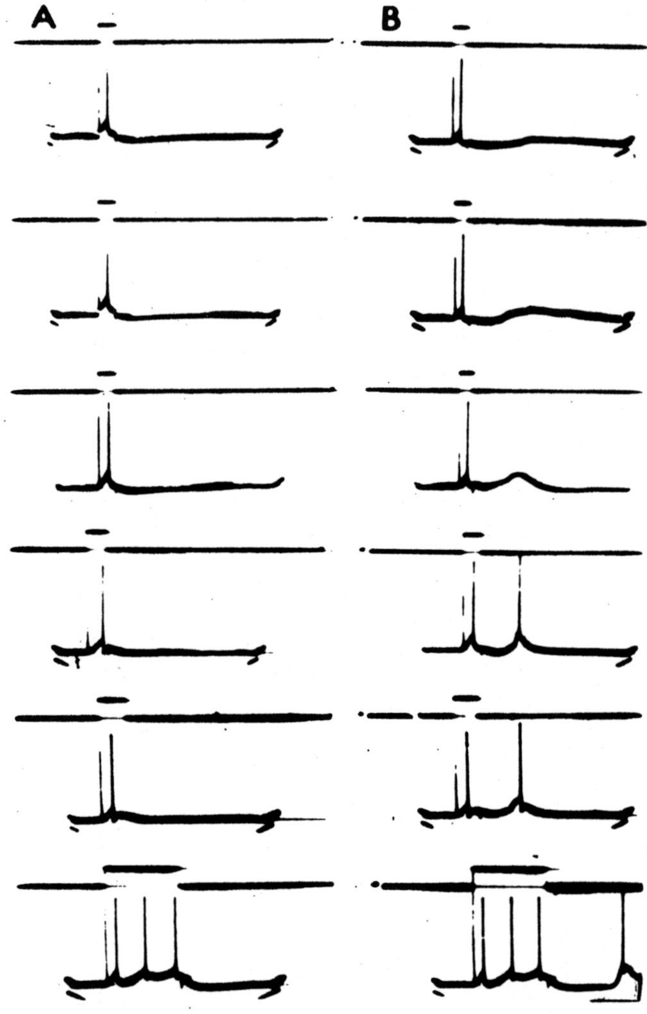

Figure 35. After-hyperpolarization curtailed by depolarizing potential as a result of passage of negative current during recording with a KCl-filled intracellular microelectrode. A was taken before and B after passage of negative current. For details see text. Upper traces show the depolarizing current through the micropipette and lower traces show responses of the cell. Calibration: 20 mV; 40 msec (Li, Okujava and Bak, 1971).

When the resulting depolarization reached the critical level, it set off spike discharges. Hence the contribution of the recurrent inhibitory synaptic influence on the development of the hyperpolarization subsequent to the action potential, must be suggested. This effect was particularly appreciable in cases when several action potentials appeared in succession. In this connection the behavour of IPSPs evoked in motor cortical cells in response to the pyramidal tract stimulation is very significant. It has been shown (Stefanis and Jasper, 1964; Humphrey, 1968; Storozhok, 1974) that their amplitude and duration depend upon the strength and, particularly, upon the number of stimuli applied to the pyramidal tract in rapid succession. In many neurons, noticeable IPSPs were evoked only in response to a brief high-frequency train of pyramidal shocks. One explanation for this fact should be that the recurrent inhibitory input attains the recorded neuron via a chain of several internuncial neurons (Serkov, 1986) (Such "breaking through" is typical of volleys in polysynaptic paths). Therefore it is of particular importance to consider the nature of postexcitatory hyperpolarization potential developing at the end of high frequency discharge of spike potentials in cortical neurons, i.e., post-burst hyperpolarization (PBH). As demonstrated in molluscan neurons (Meech, 1972), in visceral afferent neurons of rabbits (Higashi, Morita and North, 1984), in hippocampal pyramidal neurons (Alger and Nicoll, 1980; Hoston and Prince, 1980; Schwartzkroin and Stafstrom, 1980; Wong and Prince, 1981; Fournier and Crepel, 1984; Lancaster and Wheal, 1984) and in neurons of sensorimotor cortical slices of guinea pigs (Connors, Gutnick and Prince, 1982) PBH must be mainly attributed to Ca^{2+}-dependent K^+ conductance. As a matter of fact intracellular injection of either cesium ions or EGTA in our experiments (Z. G. Kokaia, Kokaia, Labakhua, Okujava, 1987a,b, 1988) strongly reduced the amplitude of PBH. As shown in Fig. 36 eleven minutes after penetrating the cell with a microelectrode filled with 0.5 M solution of Cs_2SO_4 the spontaneous

PBH became markedly depressed in comparison with the similar burst event in the beginning of the recording. In Fig. 37, where the graph of averaged values of PBHs in four neurons is presented, the depressing effect of the Cs^+ infusion upon the PBH, particularly upon its later phase, is clearly evident. A rather similar effect was observed under the influence of the EGTA injection. In Fig. 38 A the burst discharge in the PT neuron was evoked by means of depolarizing current pulse. The PBH, well pronounced in the beginning of the microelectrode penetration, became markedly depressed with total disappearance of its later phase under the influence of EGTA infusion. The same phenomenon is particularly evident in Fig 38 B where the trace is the average of eight responses. The line graph of the modifications of the

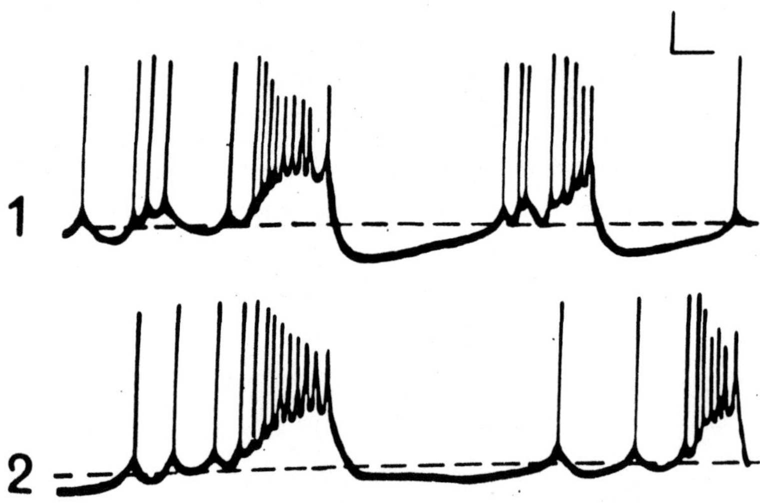

Figure 36. The influence of prolonged recording with a 0.5 M Cs_2SO_4 filled microelectrode upon the PBH in a sensorimotor cortical neuron. Curve 1 shows the beginning of the recording. Curve 2 was recorded 11 min later. Calibration: 10 mV; 50 msec (According to studies performed with Z. G. Kokaia, 1987).

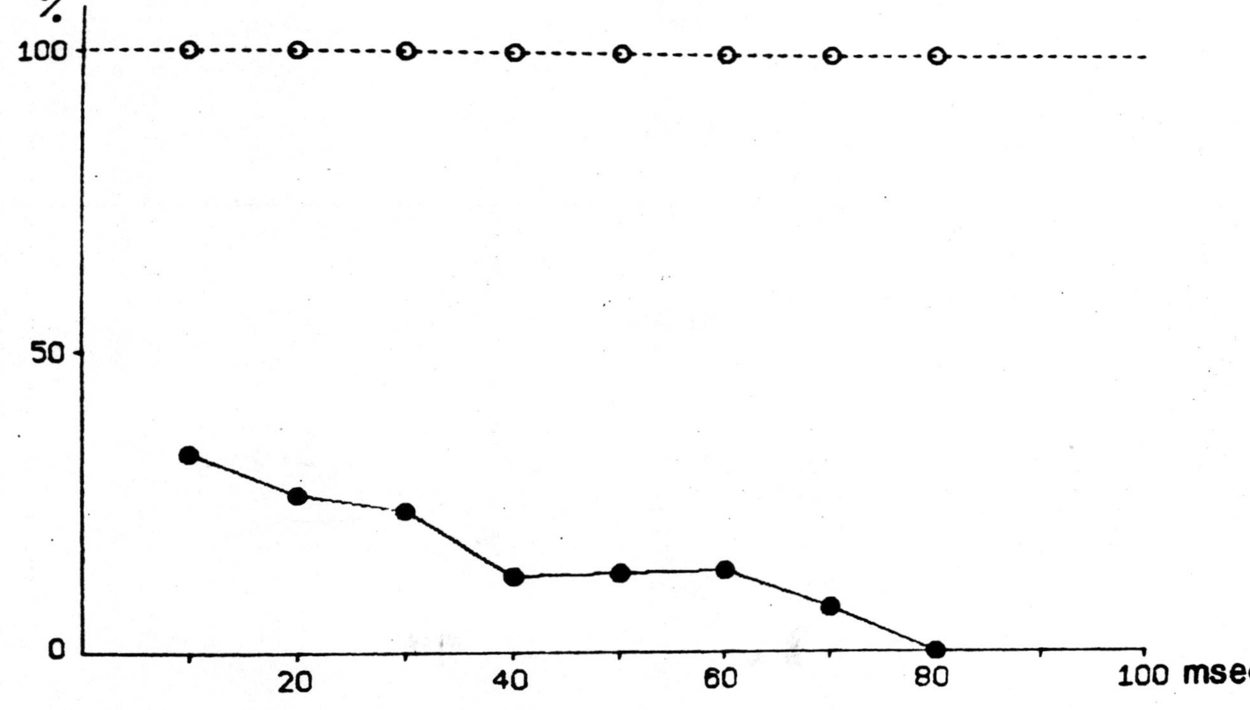

Figure 37. Influence of intracellular infusion of Cs ions upon mean values of the PBH amplitudes (closed circles) in sensorimotor cortical neurons (n=4) measured with 20 msec intervals as a % of the values of corresponding points in initial recording (open circles) (Z. G. Kokaia, Kokaia, Labakhua and Okujava, 1988).

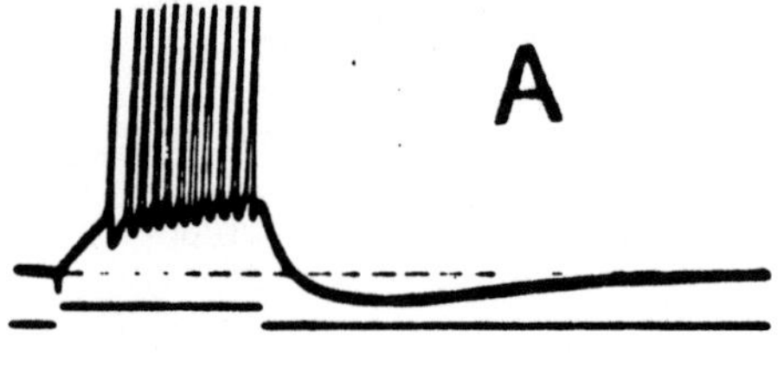

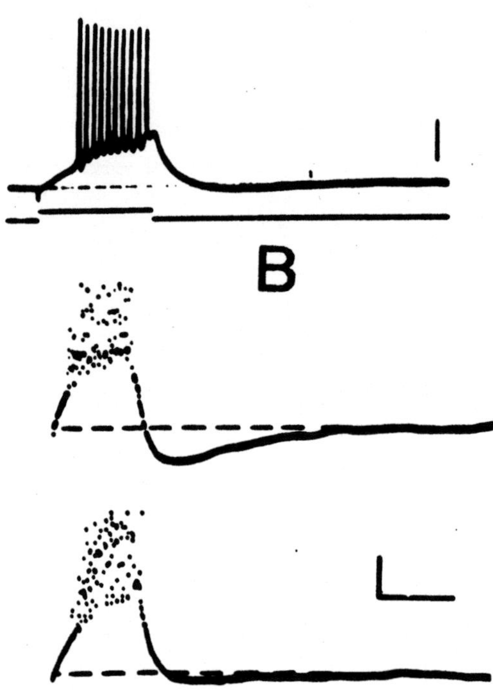

Figure 38. Effect of intracellular EGTA infusion upon the PBH of a PT neuron during recording with an EGTA-filled microelectrode. In A bursts of action potentials followed by PBH are produced as a result of passage of depolarizing current of 0.2 nA, 30 msec (shown in lower traces) in the beginning of the recording (1), and 13 min later (2). In B same events in a form of averaged curves of ten sweeps. Calibration: 10 mV; 30 msec (Z. G. Kokaia, Kokaia, Labakhua and Okujava, 1988).

PBH mean values under the influence of the EGTA injection is presented in Fig. 39, showing considerable reduction of the PBH, especially of its late part. These results clearly demonstrate the leading role of the Ca^{2+}activated K^+ conductance in generation of the PBH, particularly of its later phase The entry of the sufficient amount of Ca^{2+} ions may take place during generation of action potentials (Llinas and Hess, 1976; Schwartzkroin and Slowsky, 1977; Hagiwara and Byerly, 1981; Kostyuk and Krishtal, 1981; Connors, Gutnick and Prince, 1982; Stafstrom, Schwindt, Chubb and Crill, 1985; Franz, Galvan and Constanti, 1986). At the same time the initial part of the PBH appears to be, to a considerable extent, conditioned by chloride permeability because the early phase of the PBH becomes converted to depolarization under the influence of intracellular injection of Cl^- as demonstrated in Fig. 40. Consequently this phase of PBH must be considered to contain, at least partly, the IPSP too, as far as the chloride conductance is a specific cause of the hyperpolarizing postsynaptic potentials.

The existence of two distinct components of the PBH has been revealed also in neurons of other structures in the central nervous system (Gustaffson and Wigstrom, 1981a; Alger, 1984; Newberry and Nicoll, 1984; Greene and Haas, 1985).

In respect with recurrent inhibition of nerve cells, it must be noted that this phenomenon has been clearly demonstrated in a form of lateral inhibition at various levels of the nervous system from the peripheral receptive organs up to the neocortex.

For example, the mutual inhibition between ommatidia in the eye of Limulus, referred to by Tomita (1958) as lateral inhibition, appeared to be most likely conducted through axon collaterals (Hartline, Wagner and Ratliff, 1956; Tomita, 1958; Hartline, Ratliff and Miller, 1961).

The recurrent inhibitory pathway, presumably via interneurons, appears to be a very common arrangement in the central nervous system (cf. Eccles, 1964). In 1941 Renshow described for the first time the nonspecific inhibition of motoneurons in the spinal cord produced by action of their neighbours. As shown somewhat later (Renshaw, 1946) this inhibition, designated as "antidromic", was accompanied by repetitive discharge of interneurons in the ventral horn, lasting almost for the whole duration of inhibition. Eccles, Fatt and Koketsu (1954) discovered the cholinergic nature of the synaptic connection between recurrent motoneuron axon collaterals and these interneurons and named them Renshaw cells. The history of this system has been reviewed extensively by Eccles (1955, 1957, 1964, 1969) and also by Brooks and Wilson (1959). Granit, Pascoe and Steg (1957) and Brooks and Wilson (1959) proposed the term "recurrent" to replace the name "antidromic" inhibition.

In the brain Phillips (1959, 1961) discovered that antidromic activation of the pyramidal tract evoked large IPSPs in cortical pyramidal cells. Possibly, too, the large IPSPs observed by Lux and Klee (1962) and Klee and Lux (1962) were produced by axon collaterals of cortical cells projecting to the centrum medianum the vetro-basal complex of the thalamus and the caudate nucleus, though an alternative pathway would be via orthodromic activation of interneurons in the cortex. After Phillips, several workers have studied the cortical cells by means of intracellular techniques (for example Creutzfeldt, Lux, Nacimiento, 1964; Stefanis and Jasper, 1964a,b; Purpura and McCarthy, 1965; Creutzfeldt, Lux and Watanabe, 1966; Okujava, 1967). The existence of recurrent inhibition around the cortical pyramidal cells, so clearly established by Phillips, has since been confirmed by a considerable number of authors (e.g. Suzuki and Tukahara, 1963; Stefanis and Jasper, 1964a,b; Armstrong, 1965; Kubota, Sakata, Takahashi and Uno, 1965; Brooks and Asanuma, 1965a,b; Asanuma and Brooks, 1965; Brooks, Kameda and Nagel, 1968; Humphrey, 1968; Kameda, Nagel and Brooks, 1969). The recurrent inhibition appears to

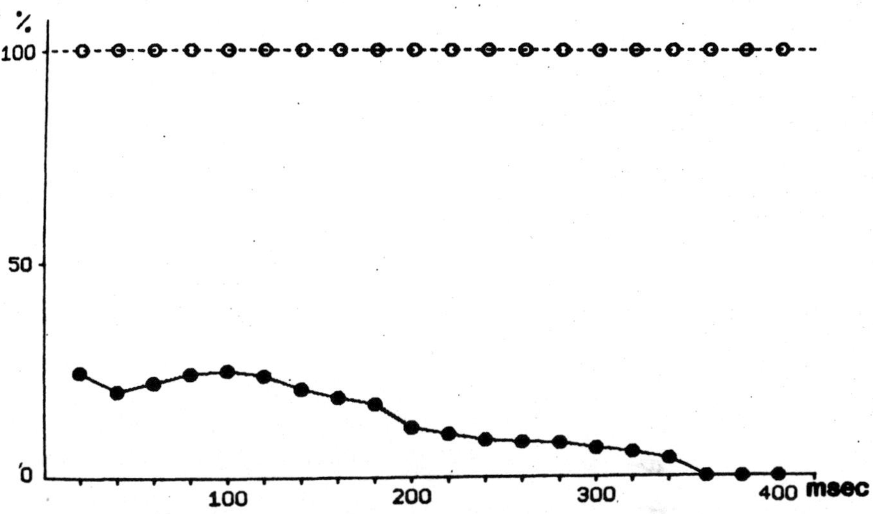

Figure 39. Influence of intracellular infusion of EGTA upon the PBHs in PT neurons (n=3). Closed circles show the amplitude of the PHBs measured with 20 msec intervals as a % of the values of corresponding points in initial recording (open circles) (Z. G. Kokaia, Kokaia, Labakhua and Okujava, 1988).

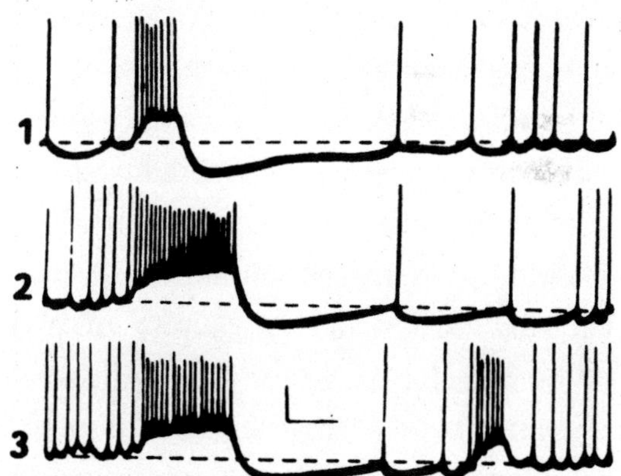

Figure 40. Differential effects of intracellular injection of chloride ions upon the early and the late components of the PBH in a PT neuron. In 1 the beginning of recording with a KCl-filled microelectrode is shown. Curve 2 and Curve 3 are the records after 32 and 33 min respectively. Calibration: 20 mV; 50 msec (Z. G. Kokaia, Kokaia, Labakhua and Okujava, 1986).

be adapted for the checking of the high rates of neuronal discharges and may well serve as a safety device preventing development of overexcitation (Phillips, 1959). According to Granit (1970) the general role of recurrent inhibition in stabilizing a discharge rate and reducing the number of firing cells to those optimally activated follows from its organization as a feedback and is thus likely to be the same wherever similar structures are found. The structural basis is supposed to be an ample supply of recurrent collaterals, which have been shown to be ubiquitous in the central nervous system (Cajal, 1911).

In addition to the stabilizing action of recurrent inhibition upon the neuronal discharge rate, its role in controlling the frequency of rhythmic waves has been suggested on several occasions, for the rhythmic waves of the lateral geniculate body (Vastola, 1959), but most notably in the generation of the theta waves of the hippocampus (Spencer and Kandel, 1962). The decisive role of recurrent IPSPs in generating rhythmic waves has been demonstrated with the ventrobasal complex of the thalamus and for the alpha rhythm of the cerebral cortex (Andersen and Eccles, 1962; Andersen, Brooks and Eccles, 1963; Eccles, 1964; Andersen and Anderson, 1968).

It must be noted that in all these experiments in which recurrent inhibition of central neurons was produced by means of antidromic stimulation of axonal bundles, convincing evidence has been obtained in favour of lateral or surround inhibition, which implies inhibition of adjacent neurons located in the vicinity of activated ones; recurrent IPSPs could be recorded in a neuron regardless of whether that particular neuron responded by an impulse. This type of recurrent inhibition, considered as lateral or surround inhibition, in the motor system presumably plays a role in localizing activity and so may be of value in distributing motor unit activity for fine movements (Wilson, 1966; Gowitzke and Milner, 1988). A similar mechanism in sensory pathways may serve to sharpen contrast (Brooks, 1959). On the other hand the results of the above-cited experiments in which AHP and PBH following activation separately of a single neuron appeared to possess definite features of postsynaptic responses, suggest the possibility of the recurrent self-inhibition of cortical neurons. Such recurrent self-inhibition with a consequent reduction of excitability in the inciting neurons can be thought of firstly as a general safety device to dampen excessive discharge.

A special feature of postexcitatory hyperpolarization in cortical neurons apparently also points to the participation of synaptic events in its origin; during direct electrical stimulation of neurons with depolarizing pulses oscillating postexcitatory hyperpolarization potentials can be revealed (Fig. 41) (Li, Okujava and Bak, 1977). The oscillations of the postexcitatory hyperpolarization potential are probably postsynaptic potentials similar to the "ripples" generated by the Renshaw cells in the spinal cord (Eccles, Fatt and Koketsu, 1954). Considerable difference in their frequencies presumably must be attributed to different characteristics of inhibitory internuncial neurons at cortical and spinal levels.

Thus it may be concluded that the postexcitatory hyperpolarization in the cortical neuron constitutes a composite phenomenon consisting, on the one hand, of after-potentials conditioned by voltage-dependent and $Ca2^+$-dependent K^+ currents, and on the other hand, of the inhibitory postsynaptic potential conditioned by synaptically activated Cl^- or Cl^- and K^+ currents, the latter component being an expression of the recurrent self-inhibition. This compound hyperpolarization must be considered as a safety mechanism counteracting the development of seizure discharges and providing the arrest of epileptic fits. However this topic will be considered in the next chapter in which the inactivation mechanisms of cortical epileptic neurons will be discussed.

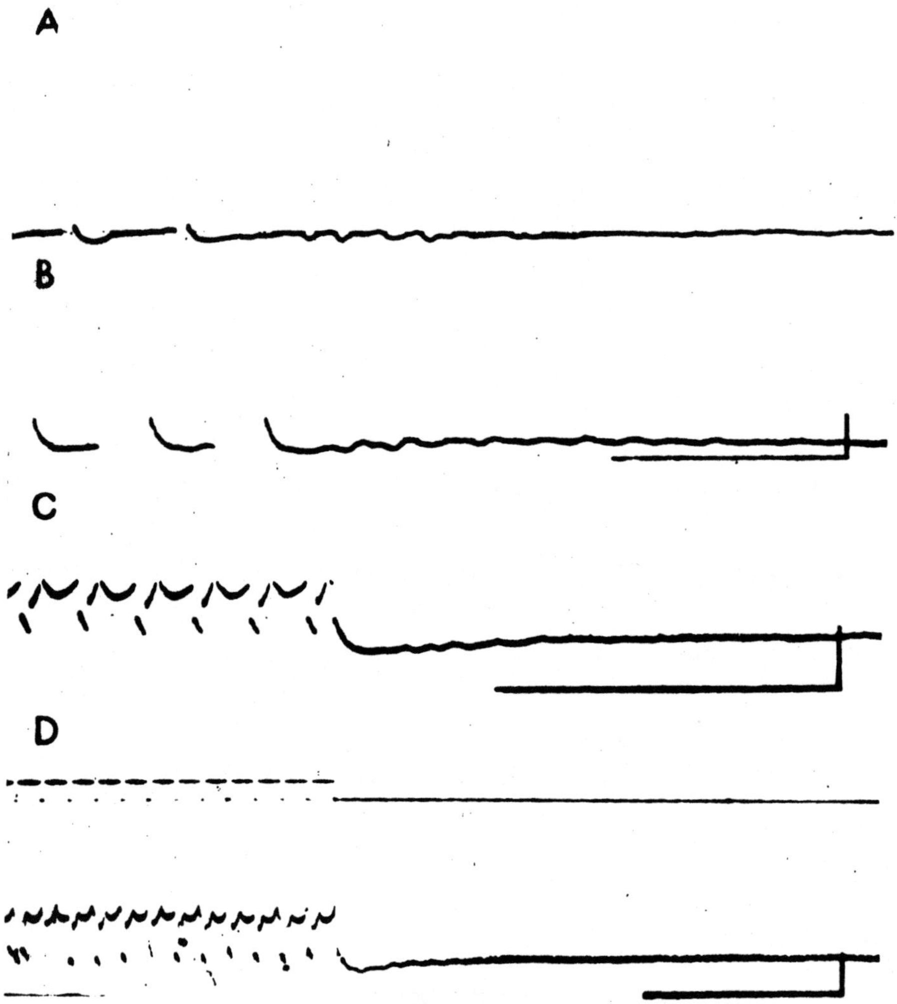

                    V.M. Okujava

A

B

C

D

Figure 41. Postexcitatory hyperpolarization potential with "oscillations". A. Responses to pulses 10 msec in duration and 0.5 x 10^{-9} A in intensity and at a frequency of 18/sec. B. Responses to pulses of 60 msec and 1.0 x 1 0^{-9} A at 15/sec. D. Responses to pulses of 40 msec and 0.8 x 10^{-9} A at 20/sec. Vartical bars represent 20 mV; horizontal bars, 400 msec (Li, Okujava and Bak, 1977).

D. The Role Of Inhibition During Neocortical Epileptic Activity

Epilepsy is characterized by abnormalities of behavior and electrical activity, which occur during recurrent convulsive and nonconvulsive seizures (ictal events), and by abnormal electrical activity, which can occur between seizures (interictal events). In spite of the pathological nature of epileptic events, their occurrences have been of great value to the neuroscientists in elucidating the form and function of the nervous system (cf. Moruzzi, 1950; Penheld and Jasper, 1954; Okujava, 1969; O'Leary and Goldring, 1976) because investigation of their mechanisms has contributed not only to the understanding of the epilepsies, but also to our understanding of neurophysiological aspects of normal brain processes. According to an apt statement of Ward, Jasper and Pope (1969) "epilepsy represents one of the most exquisite experiments of nature, and its study may provide basic insight into fundamental functions of the brain" (p.11). As a result of the study of the electrical activity in cerebral neuronal aggregates, and especially since further refinements, which permitted analysis at the individual cellular level, it has been possible to conclude that epilepsy is not a unitary process; it is likely that changes in a number of cellular processes and pathways can have the same final result in a form of abnormal, excessive electrical discharge of neurons as the main pathophysiological substrate (Ajmone, Marsan and Gumnit, 1974), confirming thus Jackson's (1931) original concept of a "sudden, excessive and rapid discharge of grey matter" as the basis of an epiletic seizure.

Because epileptic seizures and their experimental models, as well as interictal paroxysmal events, seem to correlate with extreme excitation of various cerebral structures, it is quite natural that in the majority of studies of epileptic activity excitatory phenomena predominantly attracted special attention. In fact, intracellular microelectrode studies have revealed the most characteristic feature of an epileptic neuron in a form of excessive excitatory event called the paroxysmal depolarization shift of the membrane potential (PDS) (Matsumoto, 1964; Matsumoto and Ajmone Marsan, 1964a,b).

As for the mechanisms of cessation of paroxysmal events, initially termination of seizure activity was most often assumed to be due to a metabolic exhaustion on the basis of hypoxia and to a critical acidosis and hypercapnia; the assumption seemed rather well-grounded (Jasper and Erickson, 1941; Davis, McCulloch and Roseman, 1944; Schmidt, Kety and Pennes, 1945; Davies and Remond, 1947; Wang and Sonnenschein, 1955; Ingvar, Siesjo and Hertz, 1959; Meyer and Portnoy, 1959; Ganshirt, Denneman, Schliep, Vetter and Ganshirt, 1960; Meyer, Gotoh and Tazaki, 1966). Later a number of detailed and refined studies dedicated to investigation of the metabolic state in relation with seizure termination clearly demonstrated that ictal changes of local pO_2, pCO_2 and pH did not contribute decisively to the spontaneous arrest of seizures because under present experimental conditions they did not reach levels sufficient to depress epileptic discharges (Caspers and Speckmann, 1972; Speckmann and Caspers, 1979, 1980; Speckmann, Elger and Caspers, 1980; Speckmann and Elger, 1984). On the basis of these studies the conclusion was drawn that spontaneous seizure termination is conditioned primarily by neuronal processes, namely by activation of currents through neuronal membranes—those leading to neuronal inhibition.

The role of the inhibitory process in the development, transformation, and cessation of epileptic activity has been increasingly recognized in microphysiological studies (Okujava, 1967, 1969, 1980, 1987; Prince and Widler, 1967; Prince, 1968; Matsumoto, Ayala and Gumnit, 1969; Spencer and Kandel, 1969; Ayala, Matsumoto and Gumnit, 1970; Okujava, Papitashvili and Mzhavia, 1985; Speckmann and Gutnick, 1992 etc.). According to these experimental findings, inhibitory phenomena appear to play a very important role in limitation and suppression of epileptic activity and in determining the nature and type of epileptic fits.

In most cases of epileptic activity elicited by topical application of convulsive drugs (tetanus toxin, penicillin, strychnine) to the cortex or evoked as afterdischarge by means of repetitive electrical stimulation of the cortical surface, in addition to PDSs marked hyperpolarizing potentials can be observed, which curtail each PDS restricting thus the neuronal discharge and conditioning the ultimate nature of the seizure discharge (Fig. 42). Moreover, as shown in Fig. 43, PDSs originating as interictal paroxysmal events in the cortical focus produced by topical action of penicillin, are followed by well-defined and prolonged hyperpolarizations. As far as high-frequency bursts of action potentials are triggered under the influence of PDSs, these hyperpolarization potentials might be considered as a mixture of IPSPs and intrinsic PBHs.

In many instances hyperpolarizing potentials appear to be the principal correlate of EEG paroxymal discharges. Such an example is shown in Fig. 44. However it must be emphasized that such neurons prevail in the periphery of an epileptogenic focus. This phenomenon, termed "surround" inhibition (Prince and Wilder, 1967), seems to play an important role limiting epileptic activity and preventing its generalization within the brain. Such an assumption is further substantiated by Fig. 45, where simultaneous recordings of two neurons are presented—one within the penicillin focus and the other in its immediate vicinity. Each epileptic discharge of the central neuron corresponds to profound inhibition of the peripheral one.

Pronounced inhibitory phenomena have been revealed even in experiments, in which convulsive drugs with presumably blocking action upon inhibitory synapses (tetanus toxin, strychnine) have been used for producing epileptogenic foci. The mechanisms of action of tetanus toxin had been studied mainly in respect to spinal neurons (Brooks, Curtis and Eccles, 1957; Curtis and De Groat, 1968 etc.). It has been conjectured that its convulsive effect is due to its specific blocking action upon the release of inhibitory transmitter from presynaptic terminals. Convulsive action of tetanus toxin upon the cerebral cortex was demonstrated by Carrea and Lanari (1962). Later Brooks and Asanuma (1962, 1965) demonstrated the depression of cortical recurrent inhibition under its influence. However in extracellular microelectrode studies Krnjevic, Randic and Straughan (1964, 1966) failed to reveal any appreciable changes in the inhibitory pause of unitary responses to direct cortical stimulation. This lent support to the conclusion that pharmacological properties of spinal and cortical inhibitions must be substantially different.

The following results may clarify the matter to a certain extent. Experiments were carried out in which intracellular recording in a tetanus toxin cortical focus was undertaken (Mzhavia and Okujava, 1980; Okujava, Papitashvili and Mzhavia, 1985; Okujava, 1987). Inhibitory reactions were evoked by electrical stimulation either of the cortical surface or of the pyramidal tract. Two peculiar types of neurons were discovered. In one population of neurons pronounced hyperpolarization potentials were revealed both following spontaneous PDSs (Fig. 46 A) and in response to cortical (Fig. 46 B) and pyramidal tract (Fig. 46 C) stimulations. In another population such hyperpolarizing waves appeared to be considerably reduced (Fig. 47 B, C, D). The first type of neurons was mostly found in the periphery of

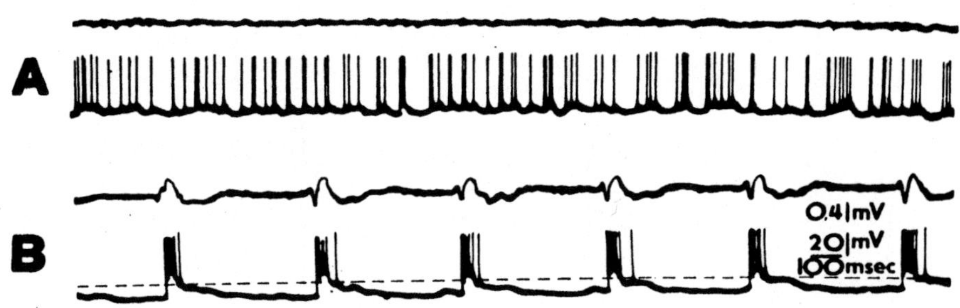

Figure 42. Electrocorticogram (upper trace) and intracellular record (lower trace) of a sensorimotor cortical neuron. A. Background activity before electrical stimulation. B. Epileptic activity following electrical stimulation of the cortical surface. Broken line represents the level of resting membrane potential (Okujava, 1987).

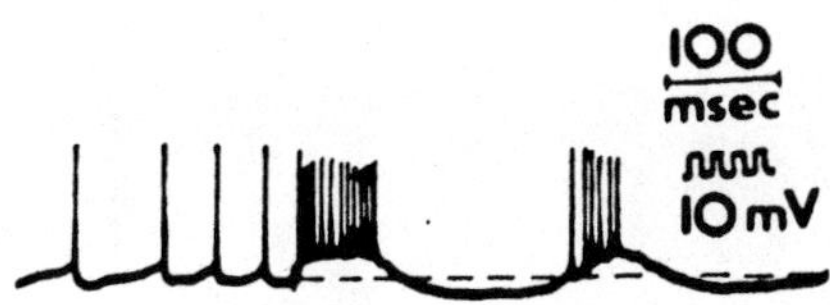

Figure 43. Intracellular record ot a sensorimotor cortical neuron during interictal paroxysmal event produced by topical application of penicillin to the cortical surface.

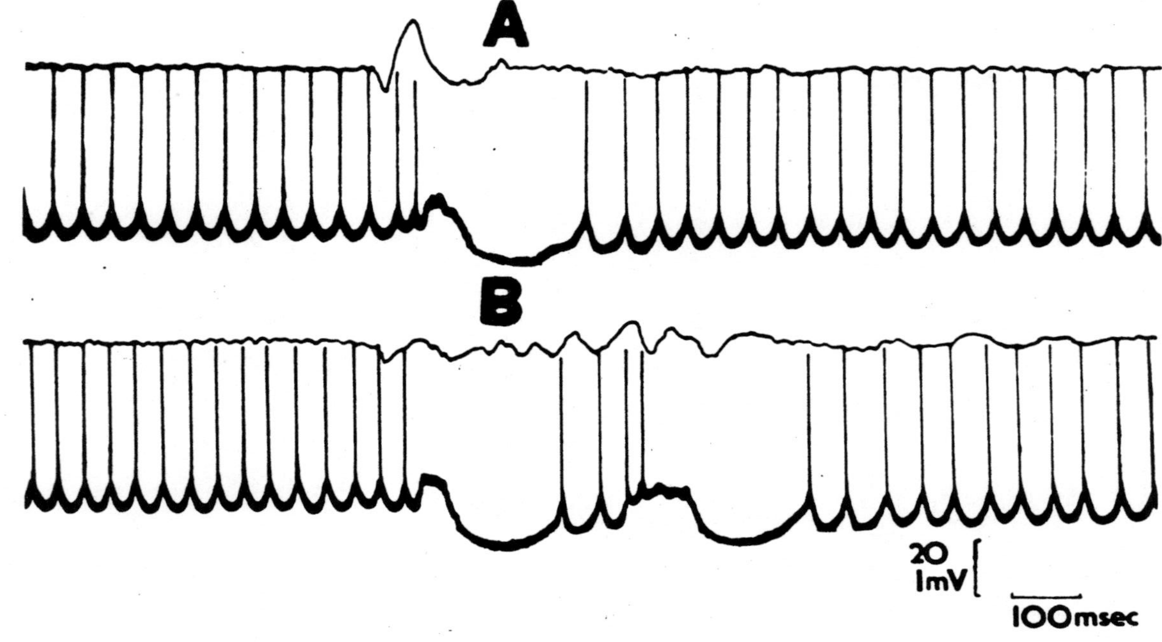

Figure 44. Two incidents (A and B) of electrocorticographic paroxysmal discharges (upper traces) and simultaneously recorded intracellular potential with marked hyperpolarization waves (lower traces) in the peripheral part of an epileptogenic focus produced by topical application of penicillin (Okujava, 1980).

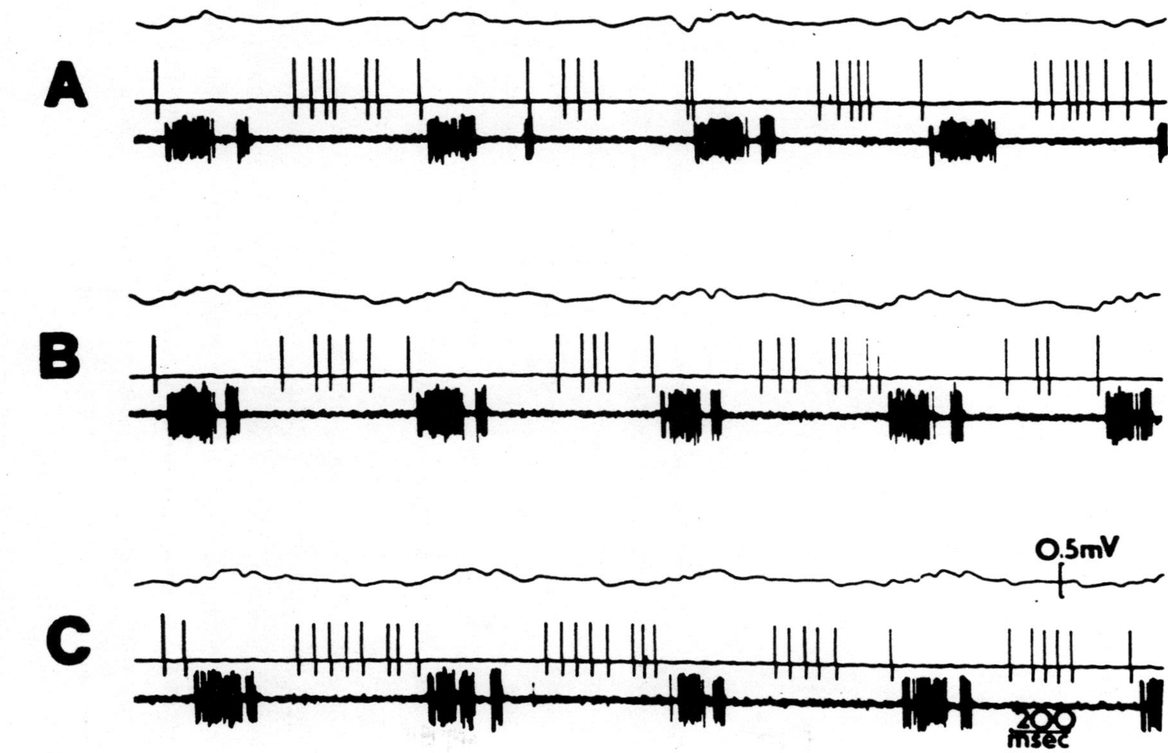

Figure 45. Simultaneous extracellular recording of two neurons: one in the center of the epileptic focus created by topical application of penicillin (lower trace), another—outside the focus (middle trace). Upper trace represents EEG (Okujava, 1987).

the focus, while the second type was regularly encountered in the central part. This testified that inhibition was blocked to a considerable extent in neurons subjected to direct and greater influence of the toxin, while the peripherally located neurons must be involved in formation of a ring of surrounding inhibition. Comparable data were obtained in experiments with strychnine. As discussed in Chapter B.2.a, in addition to surround inhibition limiting the spread of cortical epileptic activity in horizontal direction, vertical inhibition directed from superficial to deeper layers must be also considered (Elger and Speckmann, 1983a,b, 1984, 1987; Speckmann, 1986; Pockberger and Speckmann, 1989). Thus the epileptic focus appears hemispherically enveloped by an inhibitory zone. This must be an important mechanism for averting the generalization of seizure discharges.

In epileptic neurons hyperpolarizing potentials determine the character of the ictal discharge (Fig. 42). The ictal phase proper usually becomes terminated by continuous hyperpolarization (Fig. 48). Such long-lasting hyperpolarization may be the result of inhibitory synaptic action, although this might not be the only mechanism of prolonged hyperpolarization as indicated by membrane resistance measurements. As seen in Fig. 47, reduced membrane resistance gradually becomes restored up to the initial level, while hyperpolarization still remains persistent for a long time. It is likely that the activation of electrogenic pump may be the cause of such continuous hyperpolarization.

In cases when hyperpolarizing mechanisms appear insufficient for effective hyperpolarization of cortical neurons depolarizing changes, which condition their epileptic properties, increase producing inactivation by depolarization stopping thus epileptic discharges (Fig. 50). Presumably in such a case of cessation of epileptic activity, if an inactivation process occurs in a considerable number of neurons, the phenomenon of spreading depression could be initiated. The latter assumption is substantiated by the results presented in Fig. 51. In this experiment the focus of epileptic activity was produced in one point of suprasylvian gyrus by means of electrical stimulation. At a distance of 3 mm in the same gyrus no epileptic activity was recorded. In the focus epileptic activity was replaced by depression which spread following 1 minute to the point of the second recording electrode. Consequently in this case postictal depression was similar to the Leao's spreading depression (1944).

Finally, due to two such opposite mechanisms of cessation, some equilibriation of depolarizing and hyperpolarizing influences must be required for prolonged epileptic activity of a neuron. This means that neuronal "status epilepticus" must be conditioned by a certain balance between the depolarizing and hyperpolarizing mechanisms.

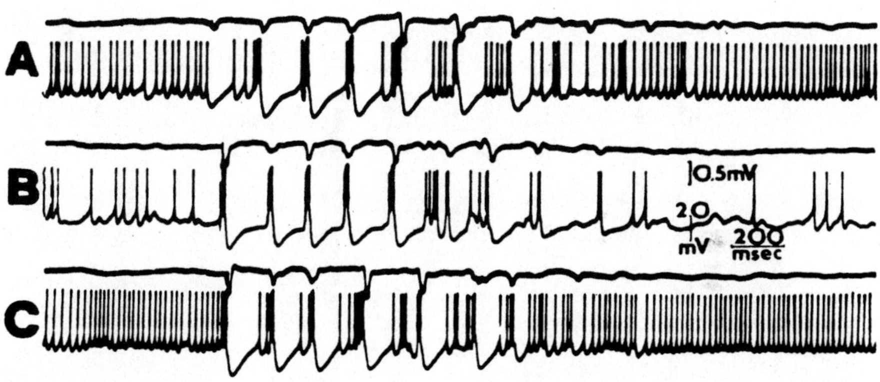

Figure 46. Fully developed hyperpolarization potentials in a neuron in the tetanus toxin cortical focus. A. Spontaneous paroxysmal discharge. B. Response to surface cortical stimulation. C. Response to pyramidal tract stimulation. Upper traces represent ECoG and lower traces—intracellular records (Okujava, Mzhavia, 1980).

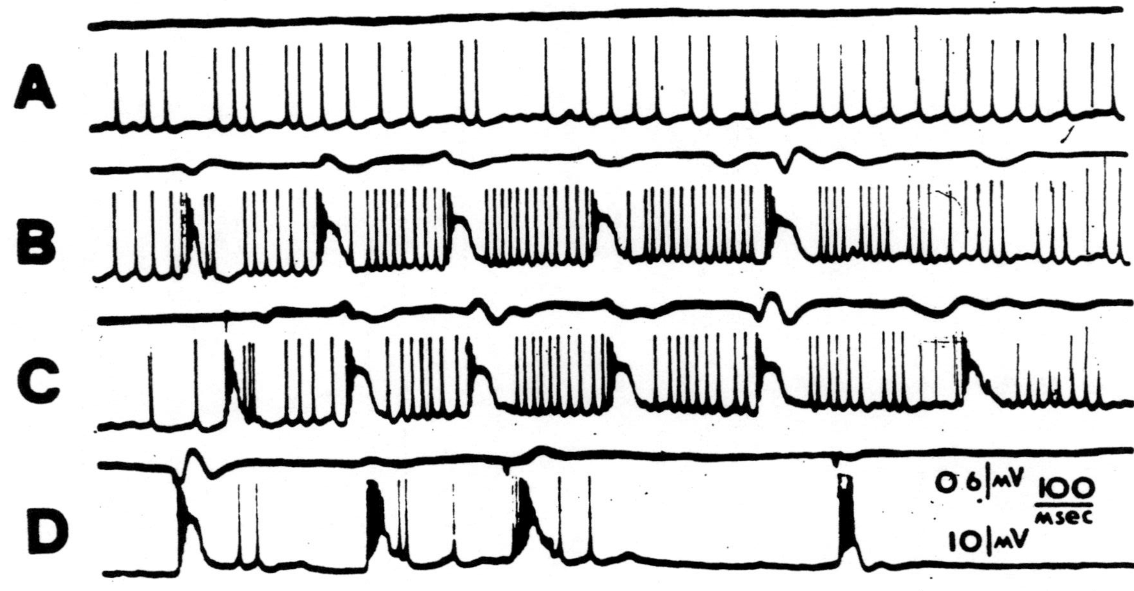

Figure 47. Reduced hyperpolarizing electrogenesis in a neuron of the tetanus toxin cortical focus. A. Activity during quiet interictal period. B. Spontaneous discharge. C. Response to cortical stimulation. D. Response to pyramidal tract stimulation. Upper traces represent ECoG and lower traces -intracellular records (Okujava, Mzhavia, 1980).

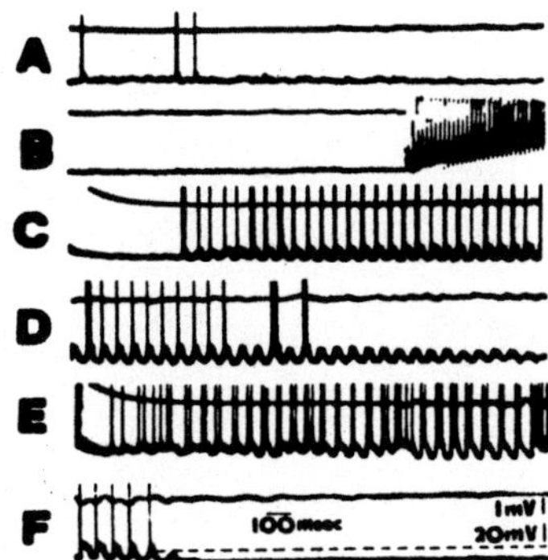

Figure 48. Intracellular recording during epileptic activity elicited by repetitive electrical stimulation of the sensorimotor cortical surface. Upper traces represent electrocorticographic and lower traces intracellular records. A. Background activity. B. Commencement of stimulation. C. Immediately following stimulation, recording continued in D. E. Immediately after repeated stimulation. Note prolonged hyperpolarization, particularly evident in F (Okujava, 1967).

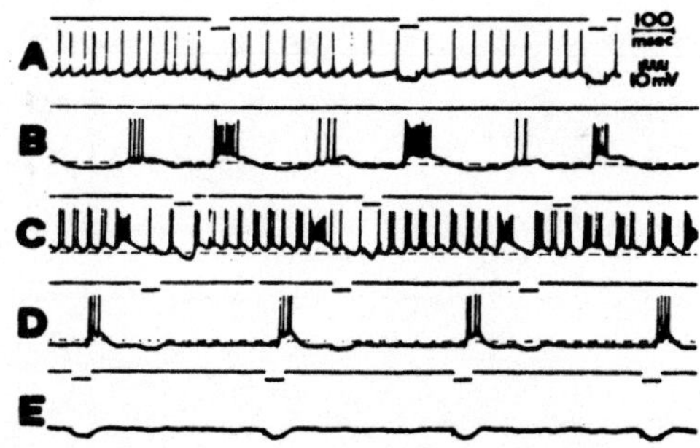

Figure 49. Neuronal membrane resistance measurements in the sensorimotor cortical penicillin focus before (A), during (B, C, D) and at the end of ictal epileptic activity. Note restoration of the resistance while hyperpolarization is continued (E). Upper trace signals intracellular injection of hyperpolarizing current pulses (Okujava, 1987).

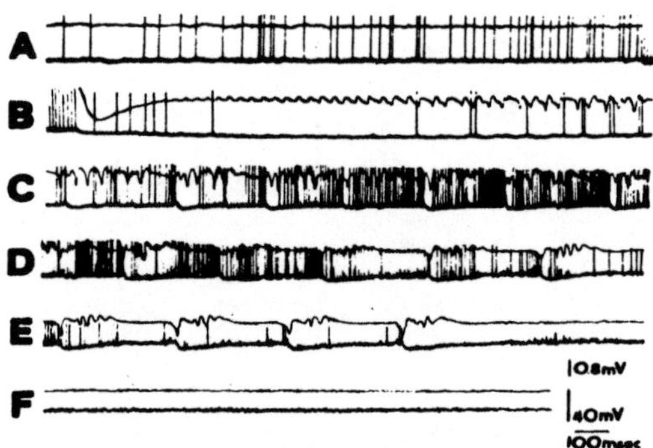

Figure 50. Cessation of epileptic activity with prolonged depolarization in a sensorimotor cortical neuron. A. Background activity before cortical repetitive electrical stimulation and the commencement of stimulation. B. Immediately following stimulation, recording continued in C, D, E, F (Okujava, 1969).

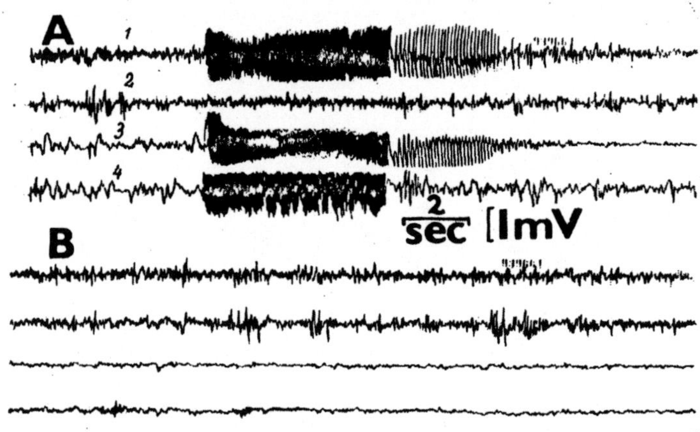

Figure 51. Spread of postictal depression. Electrocorticogram recorded in: 1) right middle suprasylvian gyrus; 2) 3 mm posterior to 1; 3) left middle suprasylvian gyrus, and 4) 3 mm posterior to 3. Repetitive electrical stimulation is applied to point 3 (note stimulation artifacts). Interval between A and B is 1 min (Okujava, 1969).

REFERENCES

Adelman, W.J., and Senft, J.P. (1958). Dynamic asymmetries in the squid axon membrane. J. Gen. Physiol. ,41 . 102-114

Adelman, W.J.,and Senft, J.P. (1966). Voltage clamp studies on the effect of cesium ion on sodium and potassium currents in the squid giant axon. J.Gen.Physiol., 50, 279-293

Ajmone Marsan, C., and Gumnit, R.J. (1974) Neurophysiological aspects of epilepsy. In: O. Magnus and A.M.Lorentz De Haas (Eds.). The Epilepsies (pp. 30-59). Amsterdam: North-Holland Publishing Co.

Akasu, T., and Koketsu, K. (1981). Voltage-clamp studies of a slow inward current in bullfrog sympathetic ganglion cells. Neurosci. Lett. , 26, 259-262.

Alger, B.E. (1984). Characteristics of a slow hyperpolarazing synaptic potential in rat hippocampal pyramidal cells in vitro. J. Neurophysiol. , 52, 892-910.

Alger, B.E. , and Nicoll, R.A. (1980). Epileptiform burst afterhyperpolarization: calcium-dependent potassium potential in hippocampal CAI pyamidal cells. Science, 210 1122-1124.

Alger, B.E., and Nicoll, R.A. (1982). Feed-foward dendritic inhibition in rat hippocampal pyramidal cells studied in vitro. J. Physiol., 328, 105-123.

Alvarez-Leefmans, F.J., and Miledi,R.(1980). Voltage sensitive calcium entry in frog motoneurons. J. Physiol., 308, 241-257.

Andersen, P., and Andersson, S.A. (1968). Physiological Basis of the Alpha Rhythm. New York: Appleton Century Crofts.

Andersen, P., Brooks, C. McC., and Eccles, J.C. (1963). Electrical responses of the ventro-basal nucleus of the thalamus. In : Progress in Brain Research, Ed. J.P. Schade. Amsterdam: Elsevier.

Andersen , P., and Eccles, J.C. (1962). Inhibitory phasing of neuronal discharge. Nature, 196, 645-647.

Andersen, P., Eccles,J.C., and Loyning, Y. (1963). Recurrent inhibition in the hippocampus with identification of the inhibitory cell and its synapses. Nature, 198, 541-542.

Andersen, P., Eccles, J.C., Loyning,Y. (1964). Location of the potentials in the somatosensory cortex of the cat. Nature, 205, 297-298.

Anwyl, R. (1977). The effect of foreign cations, pH and pharmacological agents on the ionic permeability of an excitatory glutamate-synapse. J. Physiol., 273, 389-404.

Araki, T., Ito, M., and Oscarsson, O. (1961). Anion permeability of the synaptic motoneuron. J. Physiol., 159, 410-435.

Armstrong, D.P., Mamonets, T.M. (1972). Synaptic excitation and inhibition of Betz cells by antidromic pyramidal volleys. J.Physiol., 178, 37 p.

Aetemenko, D.P., Mamonets, T.M. (1972). [Responses of neurons in posterior suprasylvian gyrus to various stimuli]. Neurophysiology, 4, 375-383 (in Russia).

Asanuma, H., and Brooks, V.B. (1965).Recurrent cortica; effects following stimulation of internal capsules. Arch. Ital. Biol., 103, 220-246.

Asanuma, H., and Okamoto, K. (1959). Unitary study of evoked activity of callosal neurons and its effect on pyramidal tract cell activity in cats. Jap. J.Physiol., 4, 437-483.

Avoli, M. (1986). Inhibitory potentials in neurons of the deep layers of the in vitro neocortical slice. Brain Res., 370, 165-170.

Avoli, M., Barra, P.F.A., Brancati, A., Deodati, M., and Vagnozzi, R. (1976). Intracellular potentials of cortical neurons and surface activity: 2. During epilepogenic discharges induced by strychnine in the rat. Boll. Soc. Ital. Biol. Sper., 52, 1676-1680.

Ayala, G.F., Matsumoto, H., and Gumnit, R.J. (1970). Excitability changes and inhibitory mechanisms in neocortiability neurons during seizures. J. Neurophysiol., 33, 73-85.

Bagust, J., Green, K.A., and Kerkut, G.A. (1981). Strychnine sensitive inhibition in the dorsal horn of mammalian spinal cord. Brain Res., 217, 425-429.

Baldissera, F., and Gustafsson, B. (1974). Firing behavior of a neuron model based on the afterhyperpolarization conductance time course and algebraical summation. Adaptation and steady state firing. Acta Physiol. Scand., 92, 27-47.

Barker, J. L., and McBurney, R. N. (1979). GABA and glycine may share the same conductance channel on cultured mammalian neurones. Nature, 277, 234-236

Barrett, E.F. and Barrett, J. N. (1976). Separation of two voltage-sensitive potassium currents and demonstration of a tetrodotoxin-resistant calcium current in frog motoneurones. J. Phys. ,255, 737-774

Barrett, E. F. , Barrett, J. N. ,and Crill, W. E. (1980) Voltage-sensitive outward currents in cat motoneurones J. Physiol . ,304, 251-276

Benzanilla, F. , and Armstrong, M. (1972). Negative conductance caused by entry of sodium and cesium ions into the potassium channels of squid axons,. J. Gen. Physical. , 60, 588-608.

Bernardi, G. Cherubini, E. Marciani , M.G. and Stanzione P.(1979). The inhibitory action of glycine on rat cortical neurones. Neurosci. Lett. ,12 , 335-338

Betz, H. (1991). Glycine receptors: heterogenous and widespread in the mammalian brain. Trends Neurosci. , 14, 458-461

Biedenbach, M. A., and Stevens, C. F.(1969). Synaptic organization of cat olfactory cortex as revealed by intracellular recording. J. Neurophysiol., 32, 204-214

Biscoe, T. J., and Curtis, D. R. (1967). Strychnine and cortical inhibition. Nature, 214, 912-915

Biscoe, T. J. Duggan, A. W. ,and Lodge, D. (1972). Antagonism between bicuculine , strychnine and picrotoxin and depressant amino-acids in the rat nervous system . Comp. gen. Pharmac., 3, 423-433.

Blaxter, T, S, and Carlen, P. L. Davies, M. F. , and Kujtan P.W. (1986) -amino-buturic acid hyperpolarizes rat hippocampal pyramidal cells through a calcium-dependent potassium conductance. J. Physiol., 373, 181-194.

Brock, L.G. Coombs, J.S., and Eccles, J.C.(1952). The recording of potentials from motoneurones with an intracellular electrode. J. Physiol. (Lond.), 117 431-460.

Brooks, V.B. (1959) Contrast and stability in the nervous system . Trans. N.Y.Acad. Sci. , 21 (2); 387-394.

Brooks, V. B. and Asanuma, H. (1962). Action of tetanus toxin in the cerebral cortex. Science, 137, 674-676.

Brooks, V,B. , and Asanuma, H. (1965b). Pharmacological studies of recurrent cortical inhibition. Amer. J. Physiol,. 208, 674-6781.

Brooks, V.B .,Curtis, D. R. , and Eccles, J.C. (1957) The action of tetanus toxin on the inhibition of motoneurones. J. Physiol.., 135, 655-672.

Brooks, V.B. Kameda, K., and Nagel, R. (1968). Recurrent inhibition in the cat's cerebral cortex. In: C. von Euler (Ed). Structure and Functions of Inhibitory Neuronal Mechanisms (pp. 327-331). New York: Pergamon

Brooks, V.B. and Wilson, V.J. (1959) Recurrent inhibition in the cat's spinal cord. J. Physiol ., 146, 380-391.

Brown, D.A., and Griffith, W.H (1983). Calcium-activated outward current in voltage-clamped hippocampal neurones of the guinea-pig. J. Physiol., 337, 287-301.

Buhrle, Ch. Ph. , and Sooohof, U. (1983). The ionic mechanism of the excitatory action of glutamate upon the membranes of motoneurones of the frog . Pflugers. Arch., 396, 154-162.

Cajal, S. R. (1911). Histologie du systeme nerveux de l'homme et de veretébrés. Madrid: Instituto Ramon y Cajal., 1952

Cantley, L. C. and Aisen, P. (1979) . The fate of cytoplasmic vanadium. Implications on (Na,K)-ATPase inhibition. J. Biol. Chem., 254, 1781-1784.

Cajal, S. R. (1911). Histologie du système nerveux de l'homme et de vertébrés. Madrid: Instituto Ramon y Cajal, 1952.

Cantley, L. C., and Aisen, P. (1979). The fate of cytoplasmic vanadium. Implications on (Na,K)-ATPase inhibition. J. Biol. Chem., 254, 1781-1784.

Cantley, L. C., Cantley, L. G., and Josephson, L. (1978). A characterization of vanadate interactions with the (Na,K)-ATPase. Mechanistic and regulatory implications. J. Biol. Chem., 253, 7361-7368.

Carrea, R., and Lanari, A. (1962). Chronic effect of tetanus toxin applied locally to the cerebral cortex of the dog. Science, 137, 342-343.

Caspers, H., and Speckmann, E.-J. (1972). Cerebral pO_2, pCO_2 and pH: Changes during convulsive activity and their significance for spontaneous arrest of seizures. Epilepsia, 13, 699-725.

Chang, H. T. (1953). Similarity in action between curare and strychnine on cortical neurons. J. Neurophysiol., 16, 221-233.

Cherubini, E., North, R. A., and Surprenant, A. (1984). Quinine blocks a calcium activated potassium conductance in mammalian enteric neurones. Br. J. Pharmacol., 83, 3-5.

Chizhenkova, R. A. (1986). [Structural and Functional Organization of the Sensorimotor Cortex]. Moscow: Nauka (in Russian).

Cohen, B. N., Fain, G. L., and Fain, M. J. (1989). GABA and glycine channels in isolated ganglion cells from the goldfish retina. J. Physiol., 418, 53-82.

Connors, B. W., Gutnick, M. J., and Prince, D. A. (1982). Electrophysiological properties of neocortical neurons in vitro. J. Neurophysiol., 48, 1302-1320.

Cook, W. A., Duncan, C. C., and Cangiano, A. (1971). Dendritic inhibition of spinal motoneurons. Proc. Internat. Union of Physiol. Sci., XXV Internat. Congr. Munich, 9, 119.

Coombs, J. S., Eccles, J. C., and Fatt, P. (1955a). The inhibitory suppression of reflex discharges from motoneurones. J. Physiol., 130, 396-413.

Coombs, J. S., Eccles, J. C., and Fatt, P. (1955b). The specific ionic conductances and the ionic movements across the motoneuronal membrane that produce the inhibitory post-synaptic potential. J. Physiol., 130, 326-373.

Crawford, J. M., and Curtis, D. R. (1964). The excitation and depression of mammalian cortical neurones by amino acids. Brit. J. Pharmacol., 23, 313-329.

Crawford, J. M., Curtis, D. R., Voorhoeve, P. E., and Wilson, V. J. (1963). Strychnine and cortical inhibition. Nature, 200, 845-846.

Creutzfeldt, O. D., and Ito, M. (1968). Functional synaptic organization of primary visual cortex neurones in the cat. Exp. Brain Res., 6, 324-352.

Creutzfeldt, O. D., Kuhnt, U., and Benevento, L. A. (1974). An intracellular analysis of visual cortical neurones to moving stimuli: Responses in a cooperative neuronal network. Exp. Brain Res., 21, 251-274.

Creutzfeldt, O. D., Lux, H. D., and Nacimiento, A. C. (1964). Intrazelluläre Reizung corticaler Nervenzellen. Pflüg. Arch. ges. Physiol., 281, 129-151.

Creutzfeldt, O. D., Lux, H. D., and Watanabe, S. (1966). Electrophysiology of cortical nerve cells. In: D. R. Purpura and M. D. Yahr (Eds.). The Thalamus (pp. 209-235). New York: Columbia University Press.

Creutzfeldt, O. D., Watanabe, S., and Lux, H. D. (1966). Relations between EEG phenomena and potentials of single cortical cells. I. Evoked responses after thalamic and epicortical stimulation. Electroenceph. clin. Neurophysiol., 20, 1-19.

Cunningham, R., and Miller, R. (1980). Electrophysiological analysis of taurine and glycine action on neurons of the mudpuppy retina. I. Intracellular recording. Brain Res., 197, 123-138.

Curtis, D. R. (1962). The depression of spinal inhibition by electrophoretically administered strychnine. Int. J. Neuropharmacol., 1, 239-250.

Curtis, D. R. (1968). Pharmacology and neurochemistry of mammalian central inhibitory processes. In: C. von Euler, S. Skoglund and U. Söderberg (Eds.). Structure and Function of Inhibitory Neuronal Mechanisms (pp. 429-456). Oxford: Pergamon Press.

Curtis, D. R., and De Groat, W. C. (1968). Tetanus toxin and spinal inhibition. Brain Res., 10, 208-212.

Curtis, D. R., Duggan, A. W., and Johnston, G. A. R. (1969). Glycine, strychnine and picrotoxin and spinal inhibition. Brain Res., 14, 759-762.

Curtis, D. R., Duggan, A. W., and Johnston, G. A. R. (1971). The specificity of strychnine as a glycine antagonist in the mammalian spinal cord. Exp. Brain Res., 12, 547-565.

Curtis, D. R., Eccles, J. C., and Lundberg, A. (1958). Intracellular recording from cells in Clarke's column. Acta Physiol. Scand., 43, 303-314.

Curtis, D. R., Felix, D., and McLellan, H. (1970). GABA and hippocampal inhibition. Br. J. Pharmacol., 40, 881-883.

Curtis, D. R., Hösli, L., and Johnston, G. A. R. (1968a). The hyperpolarization of spinal motoneurones by glycine and related amino acids. Exp. Brain Res., 5, 235-258.

Curtis, D. R., H`sli, L., and Johnston, G. A. R. (1968b). A pharmacological study of the depression of spinal neurones by glycine and related amino acids. Exp. Brain Res., 6, 1-18.

Curtis, D. R., and Lacey, G. (1994). GAGA-B receptor mediated spinal inhibition. Neuroreport, 5, 540-542.

Davidoff, R. A., Aprison, M. H., and Werman, R. (1969). The effects of strychnine on inhibition of interneurons by glycine and -aminobutyric acid. Int. J. Neuropharmacol., 8, 191-194.

Davies, P. W., and Remond, A. (1947). Oxygen consumption of the cerebral cortex of the cat during metrazol convulsions. Res. Publ. Ass. Res. Nerv. Ment. Dis., 26, 205-217.

Davis, E. W., McCulloch, W. S., and Roseman, E. (1944). Rapid changes in the oxygen tension of cerebral cortex during induced convulsions. Amer. J. Psychiat., 100, 825-829.

Deschenes, M., Paradis, M., Roy, J. P., and Steriade, M. (1984). Electrophysiology of neurons of lateral thalamic nuclei in cat: resting properties and burst discharges. J. Neurophysiol., 51, 1196-1219.

Dichter, M. A. (1980). Physiological identification of GABA as the inhibitory transmitter for mammalian cortical neurons in cell culture. Brain Res., 190, 111-121.

Dichter, M., and Spencer, A. (1969). Penicillin-induced interictal discharges from the cat hippocampus. Characteristics and topographical features. J. Neurophysiol., 32, 649-662.

Domann, R., Dorn, T., and Witte, O. W. (1989). Calcium-dependent potassium current following penicillin-induced epileptiform discharges in the hippocampal slice. Exp. Brain Res., 78, 646-648.

Domann, R., Dorn, T., and Witte, O. W. (1991). Afterpotentials following penicillin-induced paroxysmal depolarizations in hippocampal cells in vitro. Pflügers Arch., 417, 469-478.

Dreifuss, J. D., Kelly, J. S. and Krnjevic, K. (1969). Cortical inhibition and gamma-aminobutyric acid. Exp. Brain, Res., 9, 137-154.

Dubner, R., and Rutledge, L. T. (1965). Intracellular recording of the convergence of input upon neurons in cat association cortex. Exp. Neurol., 12, 349-369.

Dubois, J., and Bergman, C. (1975). Potassium accumulation in the perinodal space of frog myelinated axons. Pflügers Arch., 358, 111-124.

Dudel, J., and Kuffler, S. W. (1961). Presynaptic inhibition at the crayfish neuromuscular junction. J. Physiol., 155, 543-562.

Eccles, J. C. (1955). The central action of antidromic impulses in motor nerve fibers. Pflügers Arch., 260, 385-415.

Eccles, J. C. (1957). The Physiology of Nerve Cells. Baltimore: John Hopkins Press.

Eccles, J. C. (1961). The nature of central inhibition, Proc. Roy. Soc., B 153, 445-476.

Eccles, J. C. (1964). The Physiology of Synapses. Berlin: Springer Verlag. 1964.

Eccles, J. C. (1966). The ionic mechanisms of excitatory and inhibitory synaptic action. Ann. N. Y. Acad. Sci., 137, 473-495.

Eccles, J. C. (1969). The Inhibitory Pathways of the Central Nervous System. Springfield: Thomas.

Eccles, J. C., Fatt, P., and Koketsu, K. (1954). Cholinergic and inhibitory synapses in a pathway from motor-axon collaterals to motoneurones. J. Physiol., 126, 524-562.

Elger, C. E., and Speckmann, E.-J. (1983a). Penicillin-induced epileptic foci in the motor cortex: Vertical inhibition. Electroenceph. clin. Neurophysiol., 56, 604-622.

Elger, C. E., and Speckmann, E.-J. (1983b). Vertical inhibition in motor cortical epileptic foci and its consequences for descending neuronal activity in the spinal cord. In: E.-J. Speckmann, and C. E. Elger (Eds.). Epilepsy and Motor System (pp. 152-160). Munchen: Urban and Schwarzenberg.

Elger, C. E., and Speckmann, E. J. (1984). Die Bedeutung der vertikalen Begrenzung corticaler Krampfherde für die Krampfmotorik. In: O. Hallen, J. G. Meyer-Wahl, and J. Braun (Eds.). Epilepsie 82. Spät - und Residual - Epilepsien, Nebenwirkungen von Anticonvulsiva (pp. 328-335). Reinbek: Einhorn-Presse Verlag.

Elger, C. E., and Speckmann, E.-J. (1987). Mechanisms controlling the spatial extent of epileptic foci. In: H. G. Wieser, E.-J. Speckmann, and J. Engel, Jr. (Eds.). The Epileptic Foci. London: John Libbey.

Fanarjian, V. V., and Roitbak, A. I. (1974). [Hyperpolarization of cortical neurons during electrical stimulation of the cortex]. Dokl. Akad. Nauk SSSR, 4, 988-991 (in Russian).

Fatt, P. (1957). Sequence of events in synaptic activation of a motoneuron. J. Neurophysiol., 20, 61-80.

Folbegrova, J. (1986). The effect of vanadate on N^+, K^+-ATPase activity of the mouse cerebral cortex during bicuculine-induced seizures. Brain Res., 363, 53-61.

Fournier, E., and Crepel, F. (1984). Electrophysiological properties of in vitro hippocampal pyramidal cells from normal and staggerer mutant mice. Brain Res., 311, 87-96.

Fowler, J. C., Green, R., and Weinreich, D. (1985). Two calcium-sensitive spike afterhyperpolarizations in visceral sensory neurones of the rabbit. J. Physiol., $\underline{365}$, 59-75.

Fowler, J. C., Wonderlin, W. F., and Weinreich, D. (1985). Prostaglandins block Ca^{2+}-dependent slow spike afterhyperpolarization independent of effects of Ca^{2+} influx in visceral afferent neurons. Brain Res., $\underline{345}$, 345-349.

Frank, K., and Fuortes, M. G. F. (1955). Potentials recorded from the spinal cord with microelectrodes. J. Physiol. (Lond.), $\underline{130}$, 625-654.

Frank, K., and Fuortes, M. G. F. (1957). Presynaptic and postsynaptic inhibition of monosynaptic reflexes. Federation Proc., $\underline{16}$, 39-40.

Franz, P., Galvan, M., and Constanti, A. (1986). Calcium-dependent action potentials and associated inward currents in guinea-pig neocortical neurons in vitro. Brain Res., $\underline{355}$, 262-271.

Fuortes, M. G., and Nelson, P. G. (1963). Strychnine: its action on spinal motoneurons of cats. Science, $\underline{140}$, 806-808.

Galvn, M., Franz, P., and Constanti, A. (1985). Spontaneous inhibitory postsynaptic potentials in guinea-pig neocortex and olfactory cortex neurons. Neurosci. Lett., $\underline{57}$, 131-135.

Glavan, M., Grafe, P., and Bruggencate, G. (1982). Convulsant actions of 4-aminopyridine on the guinea-pig olfactory cortex slice. Brain Res., $\underline{241}$, 75-86.

Gänshirt, H., Denneman, W., Schliep, H., Vetter, K., and Gänshirt, L. (1960). Der Einfluss der Nähr - und Spulfunktion des Blutes auf die Aufrechterhaltung der Krampftätigkeit. Pflü gers Arch., $\underline{271}$, 185-196.

Goldberg, L. J., and Nakamura, Y. (1968). Lingually induced inhibition of masseteric motoneurones. Experimentia, $\underline{4}$, 371-373.

Gorman, A. L. F., Woolum, J. C., and Cornwall, M. C. (1982). Selectivity of the Ca^{2+}-activated and light-dependent K^{+} channels for monovalent cations. Biophys. J., $\underline{38}$, 319-322.

Gowitzke, B. A., and Milner, M. (1988). Scientific Bases of Human Movement. Baltimore: Williams and Wilkins.

Granit, R. (1970). The Basis of Motor Control. London: Academic Press.

Granit, R., Pascoe, J. E., and Steg, G. (1957). The behavior of tonic alpha and gamma motoneurones during stimulation of recurrent collaterals. J. Physiol., 138, 381-400.

Greene, R. W., and Haas, H. L. (1985). Adenosine actions on CAI pyramidal neurones in rat hippocampal slices. J. Physiol., 366, 119-127.

Gustafsson, B., and Wigstr'm, H. (1981a). Evidence for two types of afterhyperpolarization in CAI pyramidal cells in the hippocampus. Brain Res., 206, 462-468.

Gustafsson, B., and Wigström, H. (1981b). Shape of frequency-current curves in CAI pyramidal cells in the hippocampus. Brain Res., 223, 417-421.

Haberly, L. B. (1973). Unitary analysis of opossum prepyriform cortex. J. Neurophysiol., 36, 762-787.

Hagiwara, S., and Byerly, L. (1981). Calcium channel. Ann. Rev. Neurosci., 4, 69-126.

Hamerstad, J. P., and Cutler, R. W. (1972). Sodium ion movements and the spontaneous and electrically stimulated release of (3H) GABA and (14C) glutamic acid from rat cortical slices. Brain Res., 47, 410-413.

Hamill, O. P., Borman, J., and Sakmann, B. (1983). Activation of multiple-conductance state chloride channels in spinal neurones by glycine and GABA. Nature, 305, 805-808.

Hartline, H. K., Ratliff, F., and Miller, W. H. (1961). Inhibitory interaction in the retina and its significance in vision. In: E. Florey (Ed.). Nervous Inhibition (pp. 241-284). New York: Pergamon.

Hartline, H. K., Wagner, H. G., and Ratliff, F. (1956). Inhibition in the eye of Limulus. J. Gen. Physiol., 39, 651-673.

Higashi, H., Morita, K., and North, R. A. (1984). Calcium-dependent afterpotentials in visceral afferent neurones of the rabbit. J. Physiol., 355, 479-492.

Hill, R. H., Arhem, P., and Grillner, S. (1985). Ionic mechanisms of 3 types of functionally different neurons in the lampry spinal cord. Brain Res., 358, 40-52.

Hodgkin, A. L., and Huxley, A. F. (1952). A quantitative description of membrane current and its application to conduction and excitation in nerve. J. Physiol. (Lond.), 117, 500-544.

Hodgkin, A. L., and Keynes, R. D. (1955). The potassium permeability of a giant nerve fibre. J. Physiol., 128, 61-88.

Hoston, J. R., and Prince, D. A. (1980). A calcium-activated hyperpolarization follows repetitive firing in hippocampal neurons. J. Neurophysiol., 43, 409-419.

Huck, S., and Lux, H. D. (1987). Patch-clamp study of ion channels activated by GABA and glycine in cultured cerebellar neurons of the mouse. Neurosci. Lett., 79, 103-107.

Humphrey, D. R. (1968). Re-analysis of the antidromic cortical response. II. On the contribution of cell discharge and PSPs to the evoked potentials. Electroenceph. clin. Neurophysiol., 25, 421-442.

Ingvar, D. H., Lubers, D. W., and Siesjo, B. K. (1962). Normal and epileptic EEG patterns related to cortical oxygen tension in the cat. Acta Physiol. Scand., 55 210-224.

Ingvar, D. H., Siesjo, B., and Hertz, C. H. (1959). Measurement of tissue pCO_2 in the brain. Experientia, 15, 306.

Innocenti, G. M., and Manzoni, T. (1972). Response patterns of somatosensory cortical neurons to peripheral stimuli. Arch. ital. Biol., 110, 322-347.

Ito, S., and Cherubini, E. (1990). Strychnine sensitive glycine responses in immature CA3 hippocampal neurones. Europ. J. Neurosci., Suppl. 3, p. 23.

Ito, M., Kostyuk, P. G., and Oshima, T. (1962). Further study on anion permeability in cat spinal motoneurones. J. Physiol., 164, 150-156.

Iversen, L. L., Mitchell, J. F., and Srinivasan, V. (1971). The release of (-aminobutyric acid during inhibition in the cat visual cortex. J. Physiol., 212, 519-534.

Jackson, J. H. (1931). On epilepsy and epileptiform convulsions. In: J. Taylor (Ed.). Selected Writings of John Hughlings Jackson. Vol. 1. London: Hodder and Stoughton.

Jandhyala, B. S., and Hum, G. J. (1983). Physiological and pharmacological properties of vanadium. Life Sci., 33, 1325-1340.

Jasper, H. H., and Erickson, T. C. (1941). Cerebral blood flow and pH in excessive cortical discharge induced by metrazol and electrical stimulation. J. Neurophysiol., 4, 333-347.

Johnson, E. S., Roberts, M. H. T., and Straughan, D. W. (1970). Amino-acid induced depression of cortical neurones. Brit. J. Pharmacol., 38, 659-666.

Jordan, L. M., and Phillis, J. W. (1972). Acetylcholine inhibition in the intact and chronically isolated cerebral cortex. Brit. J. Pharmacol., 45, 584-595.

Jung, R. (1967). Neurophysiologie und Psychiatrie. In: H. W. Gruhle, R. Jung, W. Mayer-Gross and Muller (Eds.). Psychiatrie der Gegenwart. Forschung und Praxis. Bd 1 (pp. 325-928). Berlin: Springer-Verlag.

Kameda, K., Nagel, R., and Brooks, V. B. (1969). Some quantitative aspects of pyramidal collateral inhibition. J. Neurophysiol., 32, 540-553.

Kandel, E. R., and Spencer, W. A. (1961). Hippocampal neuron responses to selective activation of recurrent collaterals of hippocampofugal axons. Exp. Neurol., 4, 149-161.

Kandel, E. R., Spencer, W. A., and Brinley, F. J. (1961). Electrophysiology of hippocampal neurons. I. Sequential invasion and synaptic organization. J. Neurophysiol., 24, 225-242.

Katz, B. (1966). Nerve, Muscle, and Synapse, New York: McGraw-Hill.

Kazakov, V. N., and Izmestyev, V. A. (1972). [Responses of neurons of parietal association regions to stimulation of primary sensory zones]. Neurophysiology, 4, 524-530 (in Russian).

Kazakov, V. N., Izmestyev, V. A., and Perkhurova, V. D. (1972). [Neuronal and focal reactions of the parietal association cortex to various peripheral stimuli]. Neurophysiology, 4, 358-367 (in Russian).

Kehl, S. J., and McLennan, H. (1985a). An electrophysiological characterization of inhibitions and postsynaptic potentials in rat hippocampal CA3 neurons in vitro. Ext. Brain Res., 60, 299-308.

Kehl, S. J. and McLennan, H. (1985b). A Pharmacological characterization of chloride- and potassium-dependent inhibitions in the CA3 region of the rat hippocampus in vitro. Exp. Brain Res., 60, 309-317.

Kelly, J. S., and Krnjevic, K. (1969). The action of glycine on cortical neurons. Exp. Brain Res., 9, 155-163.

Kelly, J. S., Krnjevic, K., Morris, M. E., and Yim, G. K. W. (1969). Anionic permeability of cortical neurones. Exp. Brain Res., 7, 11-31.

Kernell, D. (1965). The limits of firing frequency in cat lumbo-sacral motoneurones possessing different time course of after-hyperpolarization. Acta Physiol. Scand., 65, 87-100.

Kernell, D., and Sj'holm, H. (1973). Repetitive impulse firing: comparison between neurone model based on "voltage clamp equation" and spinal motoneurones. Acta Physiol. Scand., 87, 40-56.

Kidokoro, Y., Kubota, K., Shuto, S., and Sumino, R. (1968). Reflex organization of cat masticatory muscles. J. Neurophysiol., 31, 695-708.

Klee, C. B., Crouch, T. H., and Richman, P. C. (1980). Calmoduline. Ann. Rev. Biochem., 49, 489-515.

Klee, M. R., and Lux, H.D. (1962). Intracelluläre Untersuchungen über den Einfluss hemmender Potentiale im motorischen Cortex. II. Die Wirkungen elektrischer Reizing des Nucleus caudatus. Arch. Psychiat. Nervenkr., 203, 667-689.

Klee, M. R., Shirasaki, T., Nakaye, T., Akaike, N., and Melikov, E. N. (1992). Interaction of strychnine and bicuculline with GABA- and glycine-induced chloride currents in isolate CAI neurons. In: E.-J. Speckmann and M. J. Gutnick (Eds.). Epilepsy and Inhibition (pp. 93-106). München: Urban and Schwarzenberg.

Kokaia, M. G. (1983). [Postsynaptic inhibition of cortical neurons and mechanisms of action of some anticonvulsant drugs]. Candidate's dissertation (Supervisor of studies V. M. Okujava). Inst. of Physiology, Acad. Sci. of Georgia, Tbilisi (in Russian).

Kokaia, M. G., Labakhua, T. Sh., and Okujava, V. M. (1981). [Effect of iontophoretic application of strychnine on postsynaptic reactions of pyramidal neurons in cat's sensorimotor cortex]. Proc. Acad. Sci. of Georgia (Biol. Ser.), 7, 485-492 (in Russian).

Kokaia, M. G., Labakhua, T. Sh., and Okujava, V. M. (1982). [Effect of strychnine on postsynaptic reactions of pyramidal neurons of sensorimotor cortex in the cat]. Proc. Acad. Sci. of Georgia (Biol. Ser.), 8, 232-237 (in Russian).

Kokaia, M. G., Labakhua, T. Sh., and Okujava, V. M. (1984). [Influence of strychnine on evoked potentials and postsynaptic responses of the sensorimotor cortical neurons in cat]. Neurophysiology, 16, 480-487 (in Russian).

Kokaia, Z. G. (1987). [Ionic mechanisms of hyperpolarizing electrogenesis in cortical neurons]. Candidate's dissertation. (Supervisor of studies V. M. Okujava). Inst. of Physiology, Acad. Sci. of Georgia, Tbilisi (in Russian).

Kokaia, Z. G., Kokaia, M. G., Labakhua, T. Sh., and Okujava, V. M. (1986). [Effect of intracellular injection of chloride ions on IPSP and postburst hyperpolarization in cat sensorimotor cortical neurons]. Neurophysiology, 18, 453-460 (in Russian).

Kokaia, Z. G., Kokaia, M. G., Labakhua, T. Sh., and Okujava, V. M. (1987a). [The effect of intracellular injection of Cs^+ ions on IPSP and postburst hyperpolarization of pyramidal neurons in the cat sensorimotor cortex]. Proc. Acad. Sci. of Georgia, 13, 221-225 (in Russian).

Kokaia, Z. G., Kokaia, M. G., Labakhua, T. Sh., and Okujava, V. M. (1987b). [The role of Ca^{2+}-activated K^+ conductance in hyperpolarization of pyramidal neurons in the cat sensorimotor cortex]. Proc. Acad. Sci. of Georgia, 13, 155-160 (in Russian).

Kokaia, Z. G., Kokaia, M. G., Labakhua, T. Sh., Okujava, V. M. (1988). [Participation of Ca^{2+}-dependent K^+ conductance in membrane hyperpolarization of pyramidal neurons in the cat sensorimotor cortex]. Neurophysiology, 20, 383-389 (in Russian).

Kondratyeva, I. N. (1967). [Cyclic changes of cortical neuronal activity following shortlasting stimuli. In: Modern Problems of the Nervous System Electrophysiology (pp. 148-159)]. Moscow: Nauka (in Russian).

Kostyuk, P. G., and Krishtal, O. A. (1981). [The Mechanisms of the Electrical Excitability of the Nerve Cell]. Moscow: Nauka (in Russian).

Krishtal, O. A., Osipchuk, Yu. V., and Vrublevsky, S. V. (1988). Properties of glycine-activated conductances in rat brain neurones. Neurosci. Lett., 84, 271-276.

Krivanek, J. (1981). In vivo electrical stimulation alters sensitivity of the brain (Na^+-K^+) ATPase towards inhibition by vanadate. J. Neurobiol., 12, 343-352.

Krnjevic, K. (1974a). Chemical nature of synaptic transmission in vertebrates. Physiol. Rev., 54, 418-540.

Krnjevic, K. (1974b). Synaptic transmission in the brain. In: O. Creutzfeldt and C. F. Stevens (Eds.). Handbook of EEG and Clin. Neurophysiol., v. 2 B. Basic Neurophysiology of Neuronal and Glial Potentials. Amsterdam: Elsevier.

Krnjevic, K., and Lisiewicz, A. (1972). Injection of calcium ions into spinal motoneurones. J. Physiol., 225, 363-390.

Krnjevic, K., and Phillis, J. W. (1963). Iontophoretic studies of neurones in the mammalian cerebral cortex. J. Physiol., 165, 274-304.

Krnjevic, K., Puil, E., and Werman, R. (1978). EGTA and motoneuronal afterpotentials. J. Physiol., 275, 199-223.

Krnjevic, K., Pumain, R., and Renaud, L. (1971). Effects of Ba^{2+} and tetraethylammonium on cortical neurones. J. Physiol., 215, 223-246.

Krnjevic, K., Randic, M., and Straughan, D. W. (1964). Cortical inhibition. Nature, 164, 1294-1296.

Krnjevic, K., Randic, M., and straughan, D. W. (1966a). An inhibitory process in the cerebral cortex. J. Physiol., 184, 16-48.

Krnjevic, K., Randic, M., and Straughan, D. W. (1966b). Nature of cortical inhibitory process. J. Physiol., 184, 49-77.

Krnjevic, K., Randic, M., and Straughan, D. W. (1966c). Pharmacology of cortical inhibition. J. Physiol., 184, 78-105.

Krnjevic, K., and Schwartz, S. (1967). The action of (-aminobutyric acid on cortical neurons. Exp. Brain Res., 3, 320-336.

Kubota, K., Sakata, H., Takahashi, K., and Uno, M. (1965). Location of recurrent inhibitory synapse on cat pyramidal tract cell. Proc. Jap. Acad., 41, 195-197.

Labakhua, T. Sh., Bekaya, G. L., Okujava, V. M. (1982). [Analysis of prolonged evoked potentials from the cerebral cortex]. Neurophysiology, 14, 115-121 (in Russian).

Labakhua, T. Sh., and Okujava, V. M. (1992). [Electrogenesis of the Cerebral Cortical Evoked Potentials]. Tbilisi: Metsniereba (in Russian).

Lancaster, B., and Wheal, H. V. (1984). The synaptically evoked late hyperpolarization in hippocampal CAI pyramidal cells is resistant to intracellular EGTA. Neurosci., 12, 267-275.

Larson, M. D. (1969). An analysis of the action of strychnine on the recurrent IPSP and amino-acid induced inhibitions in the cat spinal cord. Brain Res., 15, 185-200.

Leao, A. A. P. (1944). Spreading depression of activity in the cerebral cortex. J. Neurophysiol., 7, 359-390.

Leonard, J. P., and Wickelgren, W. O. (1985). Calcium spike and calcium-dependent potassium conductance in mechanosensory neurons of the lampry. J. Neurophysiol., 53, 171-182.

Levi, G., Bernardi, G., Cherubina, E., Gallo, V., Marciani, M. G., and Stanzione, P. (1982). Evidence in favour of a neurotransmitter role of glycine in the rat cerebral cortex. Brain Res., 236, 121-131.

Li, C.-L. (1959). Cortical intracellular potentials and their responses to strychnine. J. Neurophysiol., 22, 436-450.

Li, C.-L. (1963). Cortical intracellular potentials in response to thalamic stimulation. J. Cell. Comp. Physiol., 61, 165-179.

Li, C.-L., Okujava, V. M., and Bak, A. F. (1971). Some electrical measurements of cortical elements and neuronal responses to direct stimulation with particular reference to input resistance. Exp. Neurol., 31, 263-276.

Li, C.-L., Okujava, V. M., Bak, A. F. (1977). Responses of cerebro-
cortical neurons to electrical stimulation with particular reference
to epileptiform discharges. Exp. Neurol., 55, 173-186.

Llinas, R., and Hess, R. (1976). Tetrodotoxin-resistant dendritic spikes
in avian Purkinje cells. Proc. Nat. Acad. Sci. USA, 73, 2520-2523.

Llinas, R., and Sugimori, M. (1980a). Electrophysiological properties of
in vitro Purkinje cell somata in mammalian cerebellar slices. J.
Physiol., 305, 171-195.

Llinas, R., and Sugimori, M. (1980b). Electrophysiological properties of
in vitro Purkinje cell dendrites in mammalian cerebellar slices. J.
Physiol., 305, 197-213.

Logan, S. D., Lovick, T. A., West, D. C., and Wolstencroft, J. H.
(1980). Strychnine and bicuculine resistant inhibition in the dorsal
horn, on neurones inhibited by methionyl-tyrosyl-lysine. J.
Physiol., 307, P53-P54.

Lux, H. D., and Klee, M. R. (1962). Intracelluläre Untersuchungen uber
den Einfluss hemmender potentiale im motorischen Cortex. I. Die
Wirkung elektrischer Reizung unspezifischer Thalamuskerne. Arch.
Psychiat. Nervenkr., 203, 648-666.

Lux, H. D., and Pollen, D. A. (1966). Electrical constants of neurons in
the motor cortex of the cat. J. Neurophysiol., 29, 207-220.

Macdonald, R. L., and Meldrum, B. S. (1995). Principles of antiepileptic
drug action. In: R. H. Levy, R H. Mattson, and B. S. Meldrum
(Eds.). Antiepileptic Drugs (pp. 61-77). New York: Raven Press.

Mamonets, T. M. (1981). [Neuronal inhibition in association cortex
during direct cortical and transcallosal stimulation].
Neurophysiology, 13, 133-141 (in Russian).

Mangan, J. L., and Whittaker, V. P. (1966). The distribution of free
amino acids in subcellular fractions of guinea pig brain. Biochem. J.,
98, 128-137.

Marciani, M. G., Stanzione, P., Cherubini, E., and Bernardi, G. (1980). Action mechanisms of (-aminobutyric acid (GABA) and glycine on rat cortical neurones. Neurosci. Lett., 18, 169-172.

Matsumoto, H. (1964). Intracellular events during the activation of cortical epileptiform discharges. Electroenceph. clin. Neurophysiol., 17, 294-307.

Matsumoto, H., and Ajmone-Marsan, C. (1964a). Cortical cellular phenomena in experimental epilepsy: interictal manifestations. Exp. Neurol., 9, 286-304.

Matsumoto, H., and Ajmone-Marsan, C. (1964b). Cortical cellular phenomena in experimental epilepsy: ictal manifestations. Exp. Neurol., 9, 305-326.

Matsumoto, H., Ayala, G. F., and Gumnit R. J. (1969). Neuronal behavior and triggering mechanism in cortical epileptic focus. J. Neurophysiol., 32, 688-703.

Mayer, M. L., Crunelli, V., and Kemp, J. A. (1984). Lithium ions increase action potential duration of mammalian neurons. Brain Res., 293, 173-177.

McAfee, D. A., and Yorowsky, P.G. (1979). Calcium dependent potentials in mammalian sympathetic neurones. J. Physiol., 290, 507-523.

McCormick, D. A., Connors, B. W., Lighthall, J. W., and Prince, D. A. (1985). Comparative electrophysiology of pyramidal and sparsely spiny stellate neurons of the neocortex. J. Neurophysiol., 54, 782-807.

Meech, R. W. (1972). Intracellular calcium injection causes increased potassium conductance in Apolysia nerve cells. Comp. Biochem. Physiol., 42A, 493-499.

Meech, R. W. (1978). Calcium-dependent potassium activation in nervous tissues. Annu. Rev. Biophys. and Bioengineer. 7, 1-8.

Meech, R. W., and Standen, N. B. (1975). Potassium activation in Helix aspersa neurones under voltage clamp: a component mediated by calcium influx. J. Physiol. (Lond.), 249, 211-239.

Meech, R. W., and Strumwasser, F. (1970). Intracellular calcium injection activates potassium conductance in Aplysia nerve cells. Fed. Proc., 29, 834.

Meyer, J. S., Gotoh, F., and Favale, E. (1966). Cerebral metabolism during epileptic seizures in man. Electroenceph. clin. Neurophysiol., 21, 10-22.

Meyer, J. S., Gotoh, F., and Tazaki, Y. (1961). Inhibitory action of carbon dioxide and acetazolamide in seizure activity. Electroenceph. clin. Neurophysiol., 13, 762-775.

Meyer, J. S., and Portnoy, H. D. (1959). Postepileptic paralysis. Brain, 82, 162-185.

Miledi, R., Parker, I., and Schalow, G. (1980). Transmitter induced calcium entry across the postsynaptic membrane of frog end-plates measured using arsenaro III. J. Physiol., 300, 197-212.

Moruzzi, G. (1950). L'Epilepsie experimentale. Paris: Hernann.

Mountcastle, V. B., and Powell, T. P. (1959). Neural mechanisms subserving cutaneous sensibility with special reference to the role of afferent inhibition in sensory perception and discrimination. Bull. Johns Hopkins Hosp., 105, 201-232.

Mzhavia, I. A. (1982). [Investigation of neuronal inhibition and its electrocorticographic manifestation in normal and epileptic cortex]. Candidate's dissertation. (Supervisor of studies V. M. Okujava). Inst. of Clin. a. Experim. Neurology. Tbilisi (in Russian).

Mzhavia, I. A., and Okujava, V. M. (1980). [Inhibition in tetanotoxin foci in the cerebral cortex]. Proc. Adac. Sci. of Georgia (Biol. Ser.), 6, 395-401 (in Russian).

Nacimiento, A. C., Lux, H. D., and Creutzfeldt, O. D. (1964). Postsynaptische Potentiale von Nervenzellen des motorischen

Cortex nach elektrischer Reizung spezifischer und unspezifischer Thalamuskerne. Pflügers Arch., $\underline{281}$, 152-169.

Nestler, E. J., Waleas, S. I., and Greengard, P. (1984). Neuronal phosphoproteins: physiological and clinical implication. Science $\underline{225}$, 1357-1364.

Newberry, N. R., and Nicoll, R. A. (1984). A bicuculine-resistant inhibitory postsynaptic potential in rat hippocampal pyramidal cells in vitro. J. Physiol., $\underline{348}$, 239-254.

Nicoll, R. A., and Alger, B. E. (1981). Synaptic excitation may activate a calcium-dependent potassium conductance in hippocampal pyramidal cells. Science, $\underline{212}$, 957-959.

Nishi, S., and Koketsu, K. (1967). Origin of ganglionic inhibitory postsynaptic potential. Life Sci., $\underline{6}$, 2049.

Nishi, S., and Koketsu, K. (1968). Analysis of slow inhibitory postsynaptic potential of bullfrog sypathetic ganglion. J. Neurophysiol., $\underline{31}$, 717-728.

Nohmi, M., and Kuba, M. (1984). (+)-Tubocurarine blocks the Ca^{2+}-dependent K^{+}-channels of the bullfrog sympathetic ganglion cell. Brain Res., $\underline{301}$, 146-148.

Okujava, V. M. (1967a). [Description of epileptic activity in cortical neurons. In: Problems of modern neurology (pp. 303-318)]. Tbilisi, Sabchota Sakartvelo (in Russian).

Okujava, V. M. (1967b). [Principle cellular phenomena during epileptic activity in the cerebral cortex. Trans. of Inst. of clinical and Experim. Neurology (Tbilisi), $\underline{3}$, 63-106] (in Russian).

Okujava, V. M. (1969). [Basic neurophysiological mechanisms of epileptic activity]. Tbilisi: Ganatleba (in Russian).

Okujava, V. M. (1975). [Ionic mechanisms of postsynaptic potentials in the cerebral cortex]. Proc. of the XII Congr. of the I. Pavlov Physiol. Soc. of the USSR, vol. 1, p. 10. Leningrad: Nauka (in Russian).

Okujava, V. M. (1980). [The role of inhibitory processes in epileptic activity]. In: V. M. Okujava (Ed.), Neurophysiological Mechanisms of Epilepsy (pp. 51-59). Tbilisi: Metsniereba (in Russian).

Okujava, V. M. (1987). Inactivation mechanisms of cortical epileptic neurons. In: N. Chalozonitis, and M. Gola (Eds.), Inactivation of Hypersensitive Neurons (pp. 25-32). New York: Allan R. Liss.

Okujava, V. M., Mzhavia, I. A., and Goff, V. G. (1983). [Origin of slow negative potential of direct cortical response in cat]. Neurophysiology, 15, 314-320 (in Russian).

Okujava, V. M., Papitashvili, S. M., and Mzhavia, I. A. (1985). [Inhibition processes in epileptic foci of cat somatosensory cortex]. Fiziol. Zhurn., 31, 589-592 (in Russian).

O'Lague, P. H., Potter, D. O., and Furshpan, E. J. (1978). Studies on rat sympathetic neurons developing in cell culture. I. Growth characteristic and electrophysiological properties. Develop. Biol., 67, 384-403.

O'Leary, J. L., and Goldring, S. (1976). Science and Epilepsy: Neuroscience Gains in Epilepsy Research. New York: Raven Press.

Onodera, K., and Takeuchi, A. (1975). Ionic mechanism of the excitatory synaptic membrane of the crayfish neuromuscular junction. J. Physiol., 252, 295-318.

Oomura, Y., Ozaki, S., and Maeno, T. (1961). Electrical activity of a giant cell under abdominal conditions. Nature, 191, 1265-1267.

Oscarsson, O., Rosen, J., and Sulg, J. (1966). Organization of neurons in the cat cerebral cortex that are influenced from group I muscle afferents. J. Physiol., 183, 189-210.

Penfield, W., and Jasper, H. (1954). Epilepsy and Functional Anatomy of the Human Brain. Boston: Little, Brown and Co.

Phillips, C. G. (1956). Intracellular records from Betz cells in the cat. Quart. J. exp. Physiol., 41, 58-69.

Phillips, C. G. (1959). Actions of antidromic pyramidal volleys on single Betz cells in the cat. Quart. J. exp. Physiol., 44, 1-25.

Phillips, C. G. (1961). Some properties of pyramidal neurones of the motor cortex. In: G. E. W. Wolstenholme and M. O'Connor (Eds.) (pp. 4-24). The Nature of Sleep. London: J. & A. Churchill.

Phillis, J. W. (1970). The Pharmacology of Synapses. New York: Pergamon.

Phillis, J. W., and York, D. H. (1967a). Cholinergic inhibition in the cerebral cortex. Brain Res., 5, 517-520.

Phillis, J. W., and York, D. H. (1967b). Strychnine block of neural and drug induced inhibition in the cerebral cortex. Nature, 216, 922-923.

Phillis, J. W., and York, D. H. (1968a). An intracortical cholinergic inhibitory synapse. Life Sci., 7, 65-69.

Phillis, J. W., and York, D. H. (1968b). Pharmacological studies on a cholinergic inhibition in the cerebral cortex. Brain Res., 10, 297-306.

Pinsker, H., and Kandel, E. R. (1969). Synaptic activation of an electrogenic sodium pump. Science, 163, 931-935.

Pockberger, H., Speckmann, E.-J., and Walden, J. (1989). Epileptic phenomena in the neocortex: From activity of single neurons to field potentials of neuronal pools (pp. 302-310). In: E. Basar and T. H. Bullock (Eds.). Brain Dynamics. Progress and Perspectives. Berlin: Springer-Verlag.

Pollen, D. A., and Ajmone-Marsan, C. (1965). Cortical inhibitory postsynaptic potentials and strychninization. J. Neurophysiol., 28, 342-358.

Pollen, D. A., and Lux, H. D. (1966). Conductance changes during inhibitory postsynaptic potentials in normal and strychninized cortical neurons. J. Neurophysiol., 29, 369-381.

Powell, T. P., and Mountcastle, V. B. (1959). Some aspects of functional organization of the cortex of the post-central gyrus of the monkey. Bull. Johns Hopkins Hosp., 105, 133-162.

Prince, D. A. (1968). Inhibition in "epileptic" neurons. Exp. Neurol., 21, 307-321.

Prince, D. A., and Wilder, B. J. (1967). Control mechanisms in cortical epileptic foci. "Surround" inhibition. Arch. Neurol., 16, 194-202.

Pumain, R., and Heinemann, U. (1985). Stimulus- and amino acid-induced calcium and potassium changes in rat neocortex. J. Neurophysiol., 53, 1-16.

Purpura, D. P. (1973). [Intracellular study of the synaptic organization of the mammalian brain. In: Physiology and Pharmacology of Synaptic Transmission (pp. 113-145)]. Leningrad: Nauka (in Russian).

Purpura, P. D., and McCarthy, J. G. (1965). Intracellular activities and evoked potential changes during polarization of motor cortex. J. Neurophysiol., 28, 166-185.

Purpura, D. P., and Shofer, R. J. (1964). Cortical intracellular potentials during augmenting and recruiting responses: I. Effects of injected hyperpolarizing currents on evoked membrane potential changes. J. Neurophysiol., 27, 117-132.

Raabe, W., and Gumnit, R. J. (1975). Disinhibition in cat motor cortex by ammona. J. Neurophysiol., 38, 347-355.

Renaud, L. P., Kelly, J. S., and Provini, L. (1974). Synaptic inhibition in pyramidal tract neurons: Membrane potentials and conductance changes evoked by pyramidal tract and cortical stimulation. J. Neurophysiol., 37, 1144-1155.

Renshaw, B. (1941). Influence of discharge of motoneurons upon excitation of neighboring motoneurons. J. Neurophysiol., 4, 167-183.

Renshaw, B. (1946). Observations on interaction of nerve impulses in the gray matter and on the nature of central inhibition. Am. J. Physiol., 146, 443-448.

Ribaupierre, F., Goldstein, M. H., and Komshian, G. (1972). Intracellular study of the cat's primary auditory cortex. Brain Res., 48, 185-205.

Roberts, E., Chase, T. N., and Tower, D. B. (Eds.). (1976). GABA in Nervous System Function. New York: Raven Press.

Roy, J. P., Clercq, M., Steriade, M., and Deschenes, M. (1984). Electrophysiology of neurons of lateral thalamic nuclei in cat: mechanisms of long-lasting hyperpolarizations. J. Neurophysiol., 51, 1220-1235.

Salmoiraghi, G. G., and Stefanis, C. N. (1967). A critique of iontophoretic studies of central nervous system neurons. Internat. Rev. Neurobiol., 10, 1-30.

Satou, M., Mori, K., Tazawa, Y., and Takagi, S. F. (1982). Two types of postsynaptic inhibition in pyriform cortex of the rabbit: fast and slow inhibitory postsynaptic potential. J. Neurophysio., 48, 1142-1156.

Sawa, M., Maruyama, N., Kaji, S., and Nakamura, K. (1966). Action of strychnine to cortical neurons. Jap. J. Physiol., 16, 126-141.

Schmidt, C. F., Kety, S. S., and Pennes, H. H. (1945). Gaseous metabolism of the brain of the monkey. Amer. J. Physiol., 143, 33.

Schwartzkroin, P. A., and Slowsky, M. (1977). Probable calcium spikes in hippocampal neurons. Brain Res., 135, 157-161.

Schwartzkroin, P. A., and Stafstrom, C. E. (1980). Effects of EGTA on calcium-activated afterhyperpolarization in hippocampal CA3 pyramidal cells. Science, 210, 1125-1127.

Schwindt, P. C., and Crill, P. D. (1981). Voltage clamp study of cat spinal motoneurons during strychnine-induced seizures. Brain Res., 204, 226-230.

Sechenov, I. (1863). [Physiological studies of cerebral inhibitory
mechanisms of spinal reflex activity in frogs. Reprinted in Selected
Works, v. 3, Moscow, 1952] (in Russian).

Segal, M., and Barker, J. L. (1986). Rat hippocampal neurons in culture:
Ca^{2+} and Ca^{2+}-dependent K^+ conductance. J. Neurophysiol., 55,
751-766.

Serkov, F. N. (1975). [On the inhibitory processes in the cerebral
cortical neurons. In: The Mechanisms of the Cerebral Activity (pp.
402-412)]. Tbilisi: Metsniereba (in Russian).

Serkov, F. N. (1977). [Electrophysiology of the Higher Parts of
Auditory System]. Kiev: Naukova Dumka (in Russian).

Serkov, F. N. (1984). [Neuronal and synaptic mechanisms of cortical
inhibition]. Neurophysiology, 16, 394-402 (in Russian).

Serkov, F. N. (1986). [Cortical inhibition]. Kiev: Naukowa Dumka (in
Russian).

Shaban, V. M. (1972). [Responses of neurons of the anterior
suprasylvian gyrus to peripheral stimuli of various modalities].
Neurophysiology, 4, 368-374 (in Russian).

Sherrington, C. (1906). The Integrative Action of the Nervous System.
New Haven: Yale University Press, 1947.

Shirasaki, T., Klee, R. M., Nakaye, T., and Akaik, N. (1991).
Differential blockade of bicuculline and strychnine on GABA- and
glycine-induced responses in dissociated rat hippocampal pyramidal
cells. Brain Res., 561, 77-83.

Shuranova, Zh. P., and Gvozdikova, Z. M. (1971). [Neuronal responses
of the sensorimotor cortex to its direct electrical stimulation. In:
Investigation of the neuronal activity in the cerebral cortex (pp.
158-180). Moscow: Nauka (in Russian).

Sjodin, R. A. (1966). Long duration responses in squid giant axons
injected with ^{134}cesium sulfate solutions. J. Gen. Physiol., 50, 269-
278.

Skou, J. C. (1965). Enzymatic basis for active transport of Na^+ and K^+ across cell membrane. Physiol. Rev., 45, 596-607.

Skrebitsky, V. G., and Sharonova, I. N. (1972). [Synaptic phenomena during specific and unspecific inhibition of neurons in visual cortex]. Neurophysiology, 4, 349-358 (in Russian).

Skrebitsky, V. G., and Voronin, L. L. (1966). [Intracellular study of the neuronal electrical activity in the visual cortex of nonanesthetized rabbits]. Vyssh. Nerv. Deyat., 16 (5), 864-873 (in Russian).

Speckmann, E.-J. (1986). Experimentelle Epilepsieforschung. Darmstadt: Wissenschaftliche Buchgesellschaft.

Speckman, E.-J., and Caspers, H. (1979). Änderungen der Blutund Gewebegasdrucke während generalisierter Krampfaktivität im Tierexperiment. In: H. Doose, and G. Gross-Selbeck (Eds.). Epilepsie 1978. Epilepsiebedingte Hirnschäden, Psychogene Anfälle, Audiovisuelle Anfallsanalyse (pp. 26-36). Stuttgart: Georg Thieme Verlag.

Speckman, E.-J., and Caspers, H. (1980). Seizure activity in relation to changes of gas pressures in blood and tissue. In: R. Canger, F. Angeleri, and J. K. Penry (Eds.). Advances in Epileptology: XIth Epilepsy International Symposium (pp. 9-16). New York: Raven Press.

Speckmann, E.-J., and Elger, C. E. (1984). The neurophysiological basis of epileptic activity: a condensed overview. In: R. Dengen, and E. Niedermeyer (Eds). Epilepsy, Sleep and Sleep Deprivation (pp. 23-33). Amsterdam: Elsevier.

Speckmann, E.-J., Elger, C. E., and Caspers, H. (1980). Elementarpozesse der zentralnervösen Krampfaktivität. In: H. G. Mertens and H. Przuntek (Eds.). Pathologische Erregbarkeit des Nervensystems und ihre Behandlung (pp. 33-39). Berlin: Springer-Verlag.

Speckmann, E.-J., and Gutnick, M. J. (Eds.). (1992). Epilepsy and Inhibition. Munchen: Urban and Schwarzenberg.

Spencer, W. A., and Kandel, E. R. (1961). Hippocampal neuron responses to selective activation of recurrent collaterals of hippocampal axons. Exp. Neurol., 4, 149-161.

Spencer, W. A., Kandel, E. R. (1962). Hippocampal neuron responses in relation to normal and abnormal function. In: Physiologie de l'Hippocampe. Coll. int. centre national recherche sci. (Paris), 107, 71-103.

Spencer, W. A., and Kandel, E. R. (1969). Synaptic inhibition in seizures. In: H. H. Jasper, A. A. Ward Jr., and A. Pope (Eds.). Basic Mechanisms of the Epilepsies (pp. 575-603). Boston: Little, Brown and Co.

Stafstrom, C. E., Schwindt, P.C., Chubb, M. D., and Crill, W. E. (1985). Properties of persistent sodium conductance and calcium conductance of layer V neurons from cat sensorimotor cortex in vitro. J. Neurophysiol., 53, 153-170.

Stafstrom, C. E., Schwindt, P. C., Flatman, J. A., and Crill, W. E. (1984). Properties of subthreshold response and action potential recorded in layer V neurons from cat sensorimotor cortex in vitro. J. Neurophysiol., 52, 244-263.

Stafstrom, C. E., Schwindt, P. D., and Crill, W. E. (1984). Repetitive firing in layer V neurons from cat neocortex in vitro. J. Neurophysiol., 52, 264-273.

Stefanis, C. N. (1965). Further Investigation on the Electrophysiological Properties of PT Cells and the Mechanism of Interaction of Motor Function. Athens: Sotiropoulos.

Stefanis C. (1969). Interneuronal mechanisms in the cortex. In: The Interneuron (pp. 497-526). Los Angeles: Univ. of California Press.

Stafanis, C., and Jasper, H. (1964a). Intracellular microelectrode studies of antidromic responses in cortical pyramidal tract neurons. J. Neurophysiol., 27, 828-854.

Stefanis, C., and Jasper, H. (1964b). Recurrent collateral inhibition in pyramidal tract neurons. J. Neurophysiol., 27, 855-877.

Stefanis, C., and Jasper, H. (1965). Strychnine reversal of inhibitory potentials in pyramidal tract neurones. Int. J. Neuropharmacol., 4, 125-138.

Storozhuk, B. M. (1974). [Functional organization of somatic cortical neurons]. Kiev: Naukowa Dumka.

Suppes, T. (1984). A late slow depolarization unmasked in the presence of tetraethylammonium in neonatal rat sympathetic neurons in vitro. Brain Res., 293, 269-278.

Suzuki, H., and Tukahara, Y. (1963). Recurrent inhibition of the Betz cells, Jap. J. Physiol., 13, 386-398.

Takahashi, K. (1965). Slow and fast groups of pyramidal tract cells and their respective membrane properties. J. Neurophysiol., 28, 908-924.

Takata, M. (1981). Lingually induced inhibitory postsynaptic potentials in hypoglossal motoneurons after axotomy. Brain Res., 224, 165-169.

Takata, M. (1982). Inhibitory postsynaptic potentials evoked in hypoglossal motoneurons by lingual stimulation. Exp. Neurol., 75, 103-111.

Takata, M., and Ogata, K. (1980). Two components of inhibitory postsynaptic potentials evoked in hypoglossal motoneurons by lingual stimulation. Exp. Neurol., 69, 299-310.

Takeuchi, N. (1963). Effects of calcium on the conductance change of the end-plate membrane during the action of transmitter, J. Physiol., 167, 141-155.

Thompson, S. H., Prince, D. A. (1985). Activation of electrogenic sodium pump in hippocampal CA1 neurons following glutamate-induced depolarization. J. Neurophysiol., 56, 507-522.

Tokimasa, T. (1984). Calcium-dependent hyperpolarizations in bullfrog sympathetic neurons. Neurosci., 12, 929-937.

Tomita, T. (1958). Mechanism of lateral inhibition in eye of Limulus. J. Neurophysiol., 21, 419-429.

Toyoma, K., Matsunami, K., Ohno, T., and Tokashiki, S. (1974). An intracellular study of neuronal organization in the visual cortex. Exp. Brain Res., 21, 45-66.

Toyoma, K., Tokashiki, S., and Matsunami, K. (1969). Synaptic action of comissural impulses upon association efferent cells in cat visual cortex. Brain Res., 14, 418-520.

Traub, R. D. (1983). Cellular mechanisms underlying the inhibitory surround of penicillin epileptogenic foci. Brain Res., 216, 277-284.

Trautman, A., and Marty, A. (1984). Activation of Ca-dependent K channels by carbamoylcholine in rat lacrimal glands. Proc. Natl. Acad. Sci. USA, 81, 611-615.

Triller, A., Cluzeaud, F., and Korn, H. (1987). Gamma-aminobutyric acid-containing terminals can be apposed to glycine receptors at central synapses. J. Cell. Biol., 104, 947-956.

Trombley, P. Q., and Shepherd, G. M. (1994). Glycine exerts potent inhibitory actions on mammalian olfactory bulb neurons. J. Neurophysiol., 71, 761-767.

Van Dongen, P. A. M., Grillner, S., and H`kfelt, T. (1985). 5-Hydroxytryptamine (serotonine) causes a reduction in the afterhyperpolarization following the action potential in lampry motoneurons and premotor interneurons. Brain Res., 366, 320-325.

Vassilevski, N. N. (1968). [Neuronal Mechanisms of the Cerebral Cortex]. Leningrad: Meditsina (in Russian).

Vastola, E. F. (1959). After-positivity in the lateral geniculate body. J. Neurophysiol., $\underline{22}$, 258-272.

Voronin, L. L. (1967). [Postsynaptic potentials of motor cortical neurons in waking rabbits]. Fiziol. Zhurn. SSSR, $\underline{53}$ (6), 623-631 (in Russian).

Voronin, L. L. (1970). [Synaptic reactions of the sensorimotor cortical neurons to direct cortical stimulation]. Bull. Exp. Biol. Med., $\underline{70}$ (11), 15-19 (in Russian).

Voronin, L. L., and Ezrokhi, V. L. (1971). [Multisensory convergence upon motor cortical neurons in unanesthetized cats]. Neurophysiology, $\underline{3}$, 563-573 (in Russian).

Voronin, L. L., and Skrebitsky, V. G. (1967). [Extra- and intracellular studies of motor cortical neuronal responses to sound and light stimuli in waking rabbits]. Zh. Vyssh. Nerv. Deyat., $\underline{17}$, 523-533 (in Russian).

Voronin, L. L., and Tanenholtz, L. I. (1969). [Postsynaptic responses of the cat's motor cortical neurons to stimuli of various modalities]. Zh. Vyssh. Nerv. Deyat., $\underline{19}$, 71-82 (in Russian).

Wang, R. I. H., and Sonnenschein, R. R. (1955). pH of cerebral cortex during induced convulsions. J. Neurophysiol., $\underline{18}$.

Ward, A. A. Jr., Jasper, H. H., and Pope, A. (1969). Clinical and experimental challenges of the epilepsies. In: A. A. Ward, Jr., H. H. Jasper and A. Pope (Eds.). Basic Mechanisms of the Epilepsies. Boston: Little, Brown and Co.

Watanabe, S., Konishi, M., and Creutzfeldt, O. D. (1966). Postsynaptic potentials in cat's visual cortex, following electrical stimulation of afferent pathways. Exp. Brain Res., $\underline{1}$, 272-283.

Waterhouse, B. D., Moises, H. C., and Woodward, D. J. (1980). Noradrenergic modulation of somatosensory cortical neuronal responses to iontophoretically applied putative neurotransmitters. Exp. Neurol., $\underline{69}$, 30-49.

Weinstein, H., Roberts, E., and Kakefuda, T. (1963). Studies of subcellular distribution of (-aminobutyric acid and glutamic acid decarboxylase in mouse brain. Biochem. Pharmacol., 12, 503-509.

Werman, R., and Aprison, M. H. (1968). Glycine: The search for a spinal cord inhibitory transmitter. In: C. von Euler, S. Sloglund and U. Söderberg (Eds.). Structure and Function of Inhibitory Neuronal Mechanisms (pp. 473-486). Oxford: Pergamon Press.

Whitehorn, D., and Towe, A. L. (1968). Postsynaptic patterns evoked upon cells in sensorimotor cortex of cat by stimulation at the periphery. Exp. Neurol., 22, 222-242.

Willis, W. D., and Grossman, R. G. (1973). Medical Neurobiology. Saint Louis: C. V. Mosby.

Wilson, V. J. (1966). Regulation and function of Renshaw cell discharge. In: R. Granit (Ed.). Muscle Afferents and Motor Control (pp. 317-329). New York: John Wiley and Sons. 1966.

Wong, R. K. S., and Prince, D. A. (1981). Afterpotential generation in hippocampal pyramidal cells. J. Neurophysiol., 45, 86-97.

Yanovski, E. Sh. (1978). [Synaptic processes in cat's auditory cortex. In: Modern Problems of General Physiology of Excitable Structures (pp. 144-151)]. Kiev: Naukowa Dumka (in Russian).

Yanovski, E. Sh. (1986). [Parameters and peculiarities of inhibition of cat's association cortical neurons]. Fiziol. Zhurn., 32, 715-722 (in Russian).

Yoshida, S., Matsuda, Y., and Samejima, A. (1978). Tetrodotoxin-resistant sodium and calcium components of action potentials in dorsal root ganglion cells of the adult mouse. J. Neurophysiol., 41, 1096-1106.

Yoshimura, M., Polosa, C., and Nishi, S. (1986). Afterhyperpolarization mechanisms in cat sympathetic preganglionic neuron in vitro. J. Neurophysiol., 55, 1234-1246.

SUBJECT INDEX

P

PBH, 43, 46, 69, 73, 77, 81
penicillin, 81
polarization, 2, 3, 4, 6, 7, 11, 14,
    15, 19, 23, 34, 36, 37, 38, 43, 46,
    49, 50, 57, 58, 60, 63, 65, 66, 69,
    77, 80, 81, 82, 86
Postsynaptic inhibition, 1
potassium channels, 30, 38
presynaptic inhibition, 2
pyriform cortex, 15, 17, 30

S

seizure activity, 4, 20, 57, 80
seizures, 79, 80

sensorimotor cortex, 5, 10, 15, 31, 37
somatic cortex, 11
stimulation, 6, 10, 11, 13, 14, 15, 18,
    19, 22, 23, 34, 37, 38, 50, 57, 65,
    69, 76, 77, 81, 82, 87
strychnine, 17, 18, 19, 20, 22, 23,
    27, 28, 30, 57, 62, 81, 82, 86

T

thalamus, 3, 17, 74, 76

V

vanadate, 50, 52, 58
visual cortex, 13